CAPTIVITY'S COLLECTIONS

FLOWS, MIGRATIONS, AND EXCHANGES

Mart A. Stewart and Harriet Ritvo, editors

The Flows, Migrations, and Exchanges series publishes new works of environmental history that explore the cross-border movements of organisms and materials that have shaped the modern world, as well as the varied human attempts to understand, regulate, and manage these movements.

A complete list of books published in Flows, Migrations, and Exchanges is available at https://uncpress.org /series/flows-migrations-exchanges.

CAPTIVITY'S COLLECTIONS

Science,
Natural History,
and the
British Transatlantic
Slave Trade

KATHLEEN S. MURPHY

THE UNIVERSITY OF NORTH CAROLINA PRESS
Chapel Hill

© 2023 Kathleen S. Murphy

All rights reserved

Designed by Jamison Cockerham
Set in Arno, Scala Sans, Rudyard, and Fell DW Pica
by codeMantra

Cover and chapter opener art: Insect illustrations from *Biologia Centrali-Americana*, no. 51–52, vol. 2, Insecta Orthoptera (London: R. H. Porter, 1893). Courtesy of the Smithsonian Libraries and Archives. Map of the Bay of Castillos. Courtesy of the Biblioteca Nacional de España.

Manufactured in the United States of America

Portions of chapters 1 and 3 previously appeared as Kathleen S. Murphy, "Collecting Slave Traders: James Petiver, Natural History, and the British Slave Trade," *William and Mary Quarterly* 70, no. 4 (October 2013): 637–70.

Portions of chapter 3 previously appeared as Kathleen S. Murphy, "A Slaving Surgeon's Collection: The Pursuit of Natural History through the British Slave Trade to Spanish America," in *Curious Encounters: Voyaging, Collecting, and Making Knowledge in the Long Eighteenth Century*, ed. Adriana Craciun and Mary Terrall, 138–58 (Toronto: University of Toronto Press, 2019).

LIBRARY OF CONGRESS CATALOGING-IN-PUBLICATION DATA
Names: Murphy, Kathleen S., author.
Title: Captivity's collections : science, natural history, and the British transatlantic slave trade / Kathleen S. Murphy.
Other titles: Flows, migrations, and exchanges.
Description: Chapel Hill : The University of North Carolina Press, [2023] | Series: Flows, migrations, and exchanges | Includes bibliographical references and index.
Identifiers: LCCN 2023020737 | ISBN 9781469675909 (cloth) | ISBN 9781469675916 (paperback) | ISBN 9781469675923 (ebook)
Subjects: LCSH: Natural history—Great Britain—History—18th century. | Natural history—Atlantic Ocean Region—History—18th century. | Biological specimens—Collection and preservation—Great Britain—History—18th century. | Biological specimens—Collection and preservation—Atlantic Ocean Region—History—18th century. | Transatlantic slave trade—History—18th century. | Slave trade—Great Britain—History—18th century. | BISAC: HISTORY / Europe / Great Britain / General | SCIENCE / Natural History
Classification: LCC QH21.G7 M87 2023 | DDC 508.0941—dc23/eng/20230602
LC record available at https://lccn.loc.gov/2023020737

FOR
PRESTON
&
GILLY

CONTENTS

List of Illustrations ix

Acknowledgments xi

INTRODUCTION 1

1 Cannot on These Coasts Gather Amiss 19

2 Collecting for the Company 49

3 The *Asiento*'s Natural Historical Profits 71

4 Botany under the Cover of the Slave Trade 101

5 Searching for Goliath 125

6 A Flycatcher among Slave Traders 153

EPILOGUE 181

Notes 187

Bibliography 219

Index 235

ILLUSTRATIONS

FIGURES

0.1 Anatomical drawing of an armadillo *3*

1.1 Cape Coast Castle *24*

1.2 Herbaria specimens obtained by Edward Bartar *29*

1.3 *Phalaena guineensis* *42*

1.4 Herbaria specimens obtained by John Smyth *45*

3.1 Dolphins and fish observed by William Toller *78*

3.2 Map of the Bay of Castillos *80*

3.3 Kodda-pail plant (*Pistia stratiotes*) *96*

4.1 *Dorstenia* *109*

5.1 Goliath beetle (*Goliathus goliatus*) *126*

5.2 Eighteenth-century insect collectors *133*

5.3 *Gryllus caeruleus* *147*

6.1 Royal chamber of a termite nest *170*

6.2 Interior of a termitarium *172*

7.1 William Hunter's bee drawer *180*

MAPS

0.1 The slaving voyage of the *Wiltshire*, 1714–16 *xvi*

1.1 Routes of the British transatlantic slave trade, ca. 1563–1810 *18*

5.1 Search for the Goliath beetle in West Africa and West Central Africa *124*

6.1 Henry Smeathman's collecting efforts in Sierra Leone, 1771–75 *152*

TABLE

5.1 Drury's collectors in West Africa and West Central Africa by occupation *141*

ACKNOWLEDGMENTS

It is a sincere pleasure to have the opportunity to thank all the individuals and institutions whose support over the last decade made this book possible.

This project began as three tentative paragraphs added at the last minute to my dissertation. Although I did not pick the subject back up again for a few years, my colleagues and teachers at Hopkins deeply shaped it. Foremost among them was my adviser, Phil Morgan, who always modeled a generous, collaborative, and patient form of scholarship and mentorship for which I am very grateful. Reading Greg O'Malley's early drafts, especially on the *asiento*, helped me see connections that eventually led to this project, and he patiently has answered my many questions ever since. I will always be grateful for the esprit de corps that reigned among Hopkins' Early Americanists. For their friendship and wisdom over the years, I particularly thank Joe Adelman, Sarah Adelman, Toby Ditz, Katie Jorgenson Grey, Mary Ashburn Miller, Catherine Molineux, Jack Greene, James Roberts, Justin Roberts, Jessica Stern, Jessica Roney, and Molly Warsh.

I have benefited from colleagues and friends who have patiently answered questions, read drafts, and indulged conversations ranging from beetles and sloths to slavery and reparations. Richard Drayton first inspired my interest in natural history in the Atlantic World by allowing me to talk my way into his graduate seminar on the British Empire when I was an undergraduate at the University of Virginia. Stephen Behrendt encouraged the project at an early stage and was incredibly generous with his expertise and research into slave ship surgeons. Claire Gherini has been a trusted sounding board for all things relating to science and medicine in the Atlantic World. My thanks as well for their varied help along the way to Alex Borucki, Richard Coulton, James Delbourgo, Arnold Hunt, Charlie Jarvis, Peter Mancall, Miles Ogborn, Chris Parsons, Victoria Pickering, Nicholas Radburn, Carolyn Roberts, Mary Terrall, and Jane Webster.

My colleagues in the History Department, the Science, Technology & Society Program, and the College of Liberal Arts Dean's Office at Cal Poly have been wonderfully supportive. I am particularly grateful to Deans Doug Epperson and Philip Williams, and to the STS program for allowing a shared lab space to also be a place to write and retreat from administrative duties for an hour or two. An especially warm thanks to Christina Firpo and Matt Hopper for sharing their expertise and answering a steady stream of historiographical and pragmatic questions. Talented student research assistants ably assisted the project in its many stages and asked good questions along the way. My thanks to Carol Cornell, Ian Day, Jenny Freilach, Darby Leahy, Marissa Millhorn, Wendy Myren, Vanaaisha Pamnani, and Anthony Soliman. My goal was always to make this work accessible and clear to nonspecialists. The comments and questions about draft chapters raised by students enrolled in courses on the scientific revolution helped me to see where I had fallen short.

The writing of this book was generously supported by a Scholar's Award from the National Science Foundation (Award no. 1455679) and an American Council of Learned Societies Fellowship. I am grateful for the feedback and support provided by ACLS staff, NSF program officer Fred Kronz, and anonymous proposal reviewers. Research for this project was also supported by a Franklin Research Grant from the American Philosophical Society, by a Dibner History of Science Research Fellowship at the Huntington Library in San Marino, California, and by the College of Liberal Arts at Cal Poly.

My research has been facilitated by libraries, archives, and museums on both sides of the Atlantic. Museum curators and scientists kindly shared their expertise, graciously welcomed me into their spaces, and generously allowed me to view natural historical collections under their care. I am particularly indebted to Mark Carine, Jeanne Robinson, Charlie Jarvis, John Chainey, Martha Fleming, Stephen Harris, Anne Catterall, and Mark Spencer. Jeanne went above and beyond by also helping me to secure image rights. Diana Kohnke of the Sutro Library kindly sent me photos of specimens in the Petre Herbarium when the pandemic prevented travel in person. Without the able assistance of librarians and archivists at the British Library, the British Museum, the Huntington Library, the Linnean Society's Library, the National

Archives at Kew, Oxford University's Department of Plant Sciences, Oxford University's Museum of Natural History, the Royal Society of London's Library and Archives, and University of Glasgow's Archives and Special Collections, this book would not have been possible. My thanks to the Biblioteca Nacional de España and the Uppsala University Library for making materials available electronically. London's Natural History Museum is well known for the treasures displayed in its exhibit halls and carefully preserved in its storage rooms. Less celebrated are the treasures in its libraries and the efforts its staff takes to make materials accessible to humanists as well as scientists. The archival heart of this book lies in the collections of the Natural History Museum; my sincere gratitude to the museum's staff, past and present.

I am indebted to the participants at the workshops, seminars, and conferences where I tested many of the ideas in this book. I especially thank Richard Coulton and Charlie Jarvis for organizing the extraordinary Remembering James Petiver Conference at the Linnean Society of London. Thanks, too, to participants in the Writing Across Cultures Symposium held at University of California, Santa Cruz, the Slavery at the Crossroads of Medical Knowledge and Science: New Perspectives conference held at University of California, Irvine, and the Explorations, Encounters, and Circulation of Knowledge, 1600–1830 conference sponsored by the Clark Library at UCLA. The invitation to give a public (virtual) lecture at the R. W. Moriarty Science Seminar sponsored by the Carnegie Museum of Natural History was perfectly timed for helping me think through the framing of the introduction and epilogue. Members of the Cabinet of Natural History Seminar at the University of Cambridge; the Centre for Eighteenth-Century Studies at Queen Mary University of London; the Early Modern Studies Institute sponsored by the University of Southern California and the Huntington Library; the History of Science and Medicine Colloquium at Yale University; the History of Science Group at UCLA; and the History of Science Graduate Student Workshop at University of California, Santa Barbara pushed me to think about my materials in new ways. I owe particular thanks to Alex Borucki, Benjamin Breen, Mason Heberling, Sebastian Kroupa, Patrick McCray, Miles Ogborn, Carolyn Roberts, Carole Shammas, and Mary Terrall for invitations to speak at these intellectually invigorating events.

Talented editors and generous reviewers saw the promise in the project and illuminated paths that would enable me to realize it. Portions of chapters 1 and 3 previously appeared as Kathleen S. Murphy, "Collecting Slave Traders: James Petiver, Natural History, and the British Slave Trade," *William and Mary Quarterly* 7, no. 4 (October 2013): 637–70. I am deeply grateful to the *Quarterly*'s editorial staff and anonymous manuscript reviewers whose insightful critiques shaped not only the article but the book that eventually followed. Portions of chapter 3 also originally appeared in Adriana Craciun and Mary Terrall, eds., *Curious Encounters: Voyaging, Collecting, and Making Knowledge in the Long Eighteenth Century* (Toronto: University of Toronto Press, 2019). My thanks to both publishers for allowing me to repurpose that material here. My particular thanks to María García, Brandon Proia, and everyone at the University of North Carolina Press for shepherding this project. The Flows, Migrations, and Exchanges series editors, Mart Stewart and Harriet Ritvo, and the anonymous readers for UNC Press provided invaluable feedback.

My most heartfelt gratitude and debt of thanks are to my family and friends. For their wit, good humor, and friendship, I thank Regulus Allen, Jay Bettergarcia, Julie Bettergarcia, Devin Kuhn-Choi, Christina Firpo, Jane Lehr, Elizabeth Lowham, Dean Miller, Tom Trice, and Katrina Purtell. My love and thanks to my family, especially my in-laws, Charles and Wilda, the French family, and the Murphy clan. My parents always made it very clear that their unconditional support was for me and not for any project, book, or degree I pursued. For that and their love I am forever grateful. Preston believed in this project before I did; his support made it possible. He wisely persuaded me that it was okay to put aside one project unfinished in order to pursue another that I believed in. Gilly's arrival may have delayed the completion of this book, but it enriched the journey immeasurably.

CAPTIVITY'S COLLECTIONS

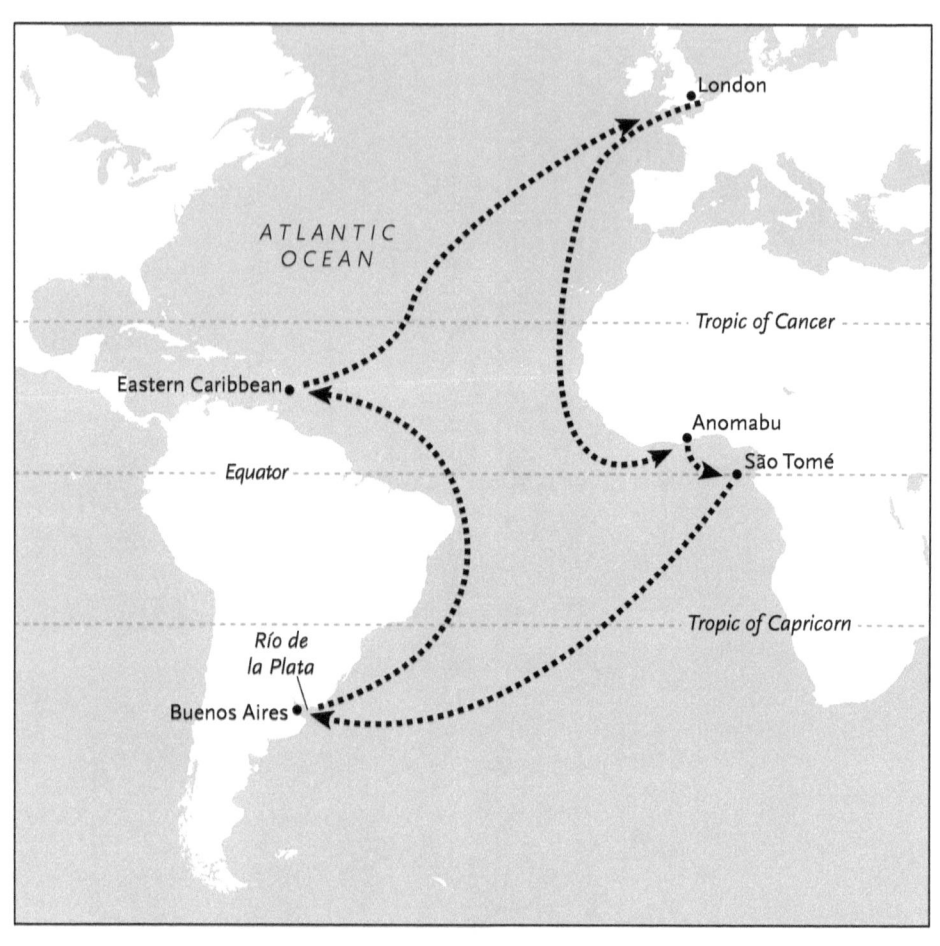

MAP 0.1. Voyage of the British slave ship *Wiltshire* to Anomabu and Buenos Aires, 1714–16. The *Wiltshire* transported nearly 300 captive African men, women, and children as well as a small collection of natural historical specimens acquired by the slave ship's surgeon, John Burnet.

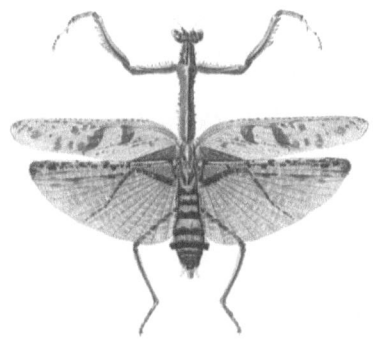

Introduction

The profits of John Burnet's first voyage as a slave ship surgeon included a sickly armadillo, an African shell, five preserved fish, three medicinal plants, and what he thought was an ostrich's egg. Burnet had departed London in 1714 as the ship surgeon on board a slaving vessel called the *Wiltshire*. During the *Wiltshire*'s two-year slaving voyage (see map 0.1), Burnet simultaneously participated in the brutal enslavement of nearly 300 African men, women, and children and collected natural historical specimens. The slave ship surgeon returned to London in the spring of 1716 with at least twenty-three specimens. Among them was "An Abortive Negroe near full grown" and three polyps "taken out of the hands of two Negroes."[1] The entangled histories of early modern science and the slave trade enabled Burnet to doubly profit from the trade, as a slave ship surgeon and as a naturalist.

Burnet collected during each stage of the *Wiltshire*'s journey: in West Africa, during the middle passage, in Spanish America, and in the eastern Caribbean on the return voyage. The *Wiltshire* was one of the first British ships to engage in the sanctioned slave trade to Spanish America. The vessel was commanded by Capt. Digory Herle and was owned by the British South Sea Company. As the ship surgeon, Burnet would have assisted Captain Herle in purchasing the 298 captive Africans whom the *Wiltshire*'s crew crowded below its decks. The *Wiltshire* then sailed for the Río de la Plata region of

Spanish America where the 247 individuals who survived the middle passage were disembarked in Buenos Aires and sold to Spanish American colonists. The *Wiltshire* returned to England on March 29, 1716.[2]

Burnet's collecting efforts began in Anomabu on the Gold Coast, where he gathered three "sucking fishes," the Guinea bitter apple, and "Tagga or earthnutts." He also acquired a worm that he noted had been "drawn by piece meal" from the leg of an individual in West Africa.[3] When the *Wiltshire* put in at São Tomé off the West African coast, Burnet collected a shell he found along the shore and the fruit of a shrub used by locals "to quench thirst in fevers."[4] The maritime nature of the next three objects in Burnet's collection suggests that they were gathered during the weeks the *Wiltshire* was at sea in the South Atlantic. These included a "fish from the other side of the Tropick of Capricorn," a young shark's jaw, and the remnants of a fish found in the belly of a shark.[5] The fish and shark in Burnet's collection were likely caught by members of the *Wiltshire*'s crew during the Atlantic crossing. The human specimens in the collection may also date to the *Wiltshire*'s middle passage. The fetus and polyps in Burnet's collection are simply described as having come from Africans. There are no further details about how Burnet obtained them or any indications about the individuals from whose bodies they were taken. However, Burnet's employment as a slave ship surgeon suggests that they were likely captive Africans on board the *Wiltshire*.

The *Wiltshire* arrived in Buenos Aires on October 4, 1715. During the weeks that the vessel was in port, Burnet must have spent at least part of his time exploring the local flora and fauna, with a particular eye out for plants that could be used as drugs or dyes. He added to his collection two plants that reportedly produced a scarlet red dye, another said to be good for treating fluxes, and a fourth known as an emetic. He also collected what he described as an ostrich egg as well as a living armadillo (fig. 0.1). Although eighteenth-century naturalists typically killed animals in the field and preserved them for transport, Burnet attempted to bring the armadillo back to London alive. The animal survived the transatlantic crossing but died a few days after the *Wiltshire* docked. Even a dead armadillo presented opportunities to advance natural knowledge. The well-known British anatomist Dr. James Douglas dissected Burnet's armadillo and reported the results in two papers presented to the Royal Society of London, the leading scientific society of its day.[6]

Burnet acquired the final two objects in his collection during the last leg of the *Wiltshire*'s journey. These specimens were the male and female West Indian "Cunny fish." Given the Atlantic's prevailing wind and current patterns, the *Wiltshire* most likely would have stopped in the eastern Caribbean to get

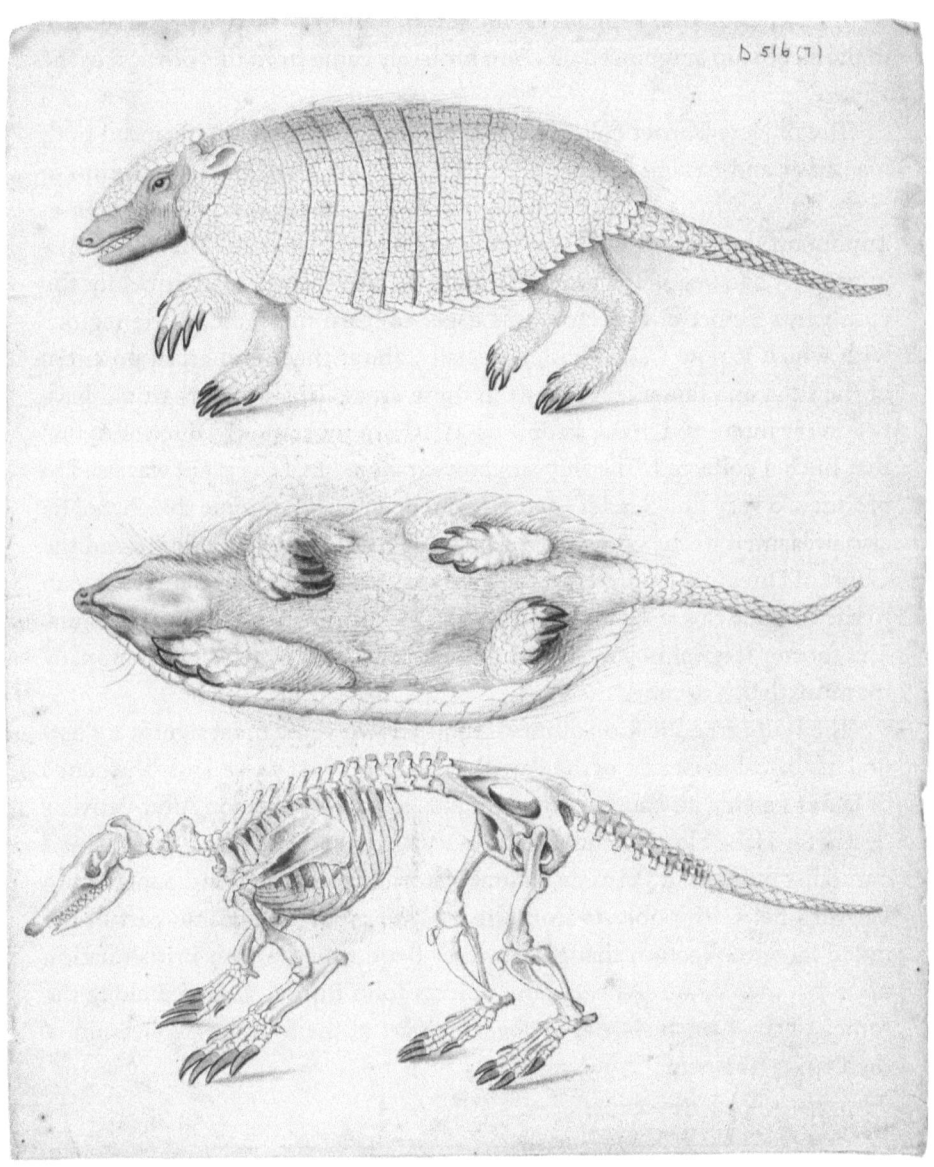

FIGURE 0.1. Anatomical drawing of an armadillo. Dr. James Douglas's drawing of the "hog in armour" is based on his dissection of the animal John Burnet acquired in Buenos Aires and transported on the slave ship *Wiltshire* in 1716. James Douglas, "The Description and Natural History of the Animal called Armadillo or the hog in armour from South America or the little American hog in Armour, by J. D.," MS Hunter D516, Archives and Special Collections, University of Glasgow, Scotland.

fresh water and other supplies for the return journey. The two preserved fish in the slave ship surgeon's collection probably came from this portion of the voyage.[7]

The objects Burnet collected while a slave ship surgeon advanced both his career and natural history. The surgeon used the *Wiltshire* collection to curry favor with his employer, the South Sea Company, which held the monopoly on the sanctioned slave trade to Spanish America. The company's monopoly had begun the year before the *Wiltshire* sailed. Consequently, the company's Court of Directors was eager to learn more about the regions with which it now traded and, especially, about the commercial potential of the flora and fauna indigenous to these areas. The directors would have been very interested in the sample of "Glashum rubrum o[f] Buenos Ayres" that Burnet collected. The ship surgeon explained that the plant was used to produce "a very fine Scarlet" dye, as evidenced by the sample dyed wool he also presented to the company. Burnet's collecting activities impressed the Court of Directors enough that the following year he was appointed surgeon to the South Sea Company's trading fort at Portobelo, in modern Panama. The former slave ship surgeon continued to work for the slaving company for more than a decade.[8]

The *Wiltshire* collection ultimately joined two of the most significant natural historical museums of the day. The collection first passed into the hands of James Petiver, an apothecary, naturalist, and avid collector. After Petiver's death, Sir Hans Sloane acquired the *Wiltshire* collection. The physician and naturalist purchased Petiver's natural historical specimens and papers from Petiver's heirs. The objects from the *Wiltshire* thereby became part of the much larger collection that Sloane later bequeathed to the British nation upon his own death. As such, the objects John Burnet gathered along the routes of the British slave trade became part of the founding collection of the British Museum.

— — — — — — — — — — — —

This book tells the story of individuals like John Burnet who acquired natural historical specimens through the exploitations of the British transatlantic slave trade. Most slaving surgeons, slave ship mariners, slave traders, and others engaged in the British transatlantic slave trade likely did not collect natural historical specimens, although it is impossible to quantify with any precision the total number of slave traders who collected. Of those British slave traders who did collect, a few may have done so out of intellectual curiosity. But the vast majority, including Burnet, collected because they stood to gain by so doing. Metropolitan British naturalists such as Sloane, Douglas, and Petiver were even greater beneficiaries of slave traders' collecting efforts. Slave traders, mariners, and surgeons acquired thousands of natural historical specimens along the routes of the British slave trade and on behalf of British naturalists. These included many natural objects that otherwise would have been difficult or impossible for British naturalists to obtain. The specimens that collecting slave traders acquired shaped the production of natural knowledge in the eighteenth century and beyond.

Captivity's Collections focuses on the eighteenth century, when the British dominated the transatlantic slave trade. English merchants replaced the Dutch and the Portuguese as the leading shippers of enslaved Africans by 1670.[9] Annual cargoes of captive Africans on British ships rose sixfold over the century that followed. *Captivity's Collections* begins in the late seventeenth century, when the Royal African Company held the monopoly on English trade to West Africa. The company's loss of its monopoly at the turn of the eighteenth century inaugurated a dramatic expansion in Britain's participation in the slave trade. More British slaving voyages meant more opportunities for British naturalists to exploit the trade in order to obtain natural historical specimens. *Captivity's Collections* ends in the 1780s, just as the abolition movement to end the slave trade was taking root in Britain.

Britain's position as the dominant enslaving power of the eighteenth century facilitated British naturalists' acquisition of specimens. British naturalists exploited the routes, personnel, and infrastructure of the transatlantic slave trade to collect seeds, shells, preserved animals, pressed plants, fossils, and other *naturalia* from around the Atlantic basin. British naturalists in the eighteenth century rarely acknowledged their reliance upon the slave trade. They would have found it no more worthy of comment than their reliance on any other long-distance trade. Few white Britons questioned slavery or the slave trade for much of the eighteenth century. This slowly changed during the century's last decades as the movement to end the slave trade gained strength in Britain. A handful of British naturalists in the 1770s started to question natural history's reliance on the slave trade. Yet for most of the period discussed in *Captivity's Collections*, the connections between the slave trade and natural history largely passed unquestioned among British naturalists.[10]

British naturalists enlisted slaving mariners and traders to collect on their behalf in each of the major regions of the British transatlantic slave trade: West Africa, the Caribbean, Spanish America, and British North America.

Britons engaged in the slave trade gathered specimens from transatlantic slaving vessels; at British slaving factories in West Africa; in British American ports where captive men, women, and children were forced to disembark; and, during the early eighteenth century, near South Sea Company factories in Spanish territories. *Captivity's Collections* traces the itineraries traveled by African and American specimens in British natural history museums to reveal how early modern collecting practices relied on the inhuman commerce in African captives.

Collecting was central to the practice of natural history and to scientific knowledge-making in the eighteenth century. Eighteenth-century naturalists aspired to an encyclopedic knowledge of all the world's natural productions. Objects collected along the routes of the British slave trade joined the herbariums, botanic gardens, cabinets, and museums that formed the basis of natural historical practice in the early modern era. Collections of specimens enabled European naturalists to study organisms from around the globe. Naturalists often traded, gifted, loaned, and published images of specimens. They thereby amplified the potential impact an individual specimen might have on the production of natural knowledge. Eighteenth-century natural history is frequently associated with the period's great systematizers, including Carolus Linnaeus and Georges-Louis Leclerc de Buffon, who undertook their influential taxonomic projects during this period. Eighteenth-century naturalists also were engaged in chemical analyses, dissections, microscopy, and other experimental methods.[11] Ready access to specimens from around the world was central to both taxonomy and experimental study.

Natural history was the "big science" of the early modern era.[12] Similar to physics in the twentieth century, the pursuit of natural history in the eighteenth century required extensive financial and human capital. European collections of specimens from around the globe represented the labor of thousands of individuals who observed, identified, collected, preserved, and transported *naturalia*. These individuals included enslaved and free Africans, Native Americans, soldiers, sailors, slave traders, colonizers, and colonized peoples. To paraphrase historian James Delbourgo, building and maintaining the polycentric social networks through which naturalists acquired specimens was not simply the context for naturalists' work but constituted a major aspect of the work itself.[13] The vast collections acquired by metropolitan naturalists such as Petiver and Sloane were built upon the efforts, both free and coerced, of a range of social actors. Surviving historical sources acknowledge only a small fraction of the collectors whose labor and knowledges were materially embodied in the seeds, specimens, and observations upon which the practice of natural history depended.

The ambiguity of the word "collector" can further obscure the social fabric of early modern natural history. "Collector" can refer to individuals like Sloane who amassed significant collections of natural specimens, akin to how the term is applied to describe individuals who acquired extensive collections of fine art. In contrast, an older tradition within the historiography of science used "collector" (often "mere collector") to describe those working in the field rather than naturalists working in European centers. Yet the dichotomy created by this use of the term quickly breaks down upon closer examination. An individual, for example, might be both a field collector and a metropolitan naturalist. Finally, specimen labels and institutional provenance records tend to identify one individual as the object's collector even though collecting parties of one were rare. Most early modern European collectors relied on local individuals as guides, porters, hunters, technicians, and informants. Further, the role played by the individual identified as the collector on a specimen label can vary greatly. It may reference an individual who donated or sold the specimen, an intermediary or agent who facilitated a specimen's acquisition, or an individual involved in physically gathering it in the field. Eighteenth-century specimen labels and provenance records rarely acknowledge the contributions of lower-status individuals or those not of European descent.[14] *Captivity's Collections* employs the term "collector" expansively to describe individuals involved in selecting and obtaining plants, animals, shells, fossils, and other objects of natural historical study. It therefore includes individuals ranging from anonymized West African collectors and slaving mariners to traveling naturalists and cabinet collectors who never left Europe.

Commercial vessels brought thousands of previously unknown plants, animals, fossils, shells, and other *naturalia* into Europe during the early modern period. The infrastructure of long-distance trade, including its routes, personnel, and systems of trust and credit, made possible the extraction and transportation of flora, fauna, and natural knowledge. Trading companies transported specimens and natural observations as well as employed the individuals who collected and composed them. Early modern commercial networks shaped issues of epistemological credibility and the processes by which new matters of fact were created.[15] Recent scholarship has shown how long-distance commerce in the early modern world shaped natural inquiries. Yet relatively little of this work has specifically focused on the trade that most defined the Atlantic World—and resulted in the forced migration of over 12.5 million Africans.[16] How did the infrastructures of the transatlantic slave trade, including its geographies, structures, trading companies, and personnel, enable natural historical collecting and thereby the production of natural knowledge? Through its focus on the British slave trade, *Captivity's Collections* contributes to our understanding of the ways in which commerce shaped the production of natural knowledge in the early modern period.

The vessels and routes of the British transatlantic slave trade transported people, specimens, and observations essential to the pursuit of natural history in the eighteenth century. However, the entangled histories of natural history and enslaving go beyond transportation. The geographies, commercial priorities, trading practices, and maritime labor of the British transatlantic slave trade determined to a large degree the social and material practices of natural historical collecting along its routes. Among the unique aspects of a slave ship compared to other commercial vessels traversing the Atlantic Ocean was that its crew typically included a ship surgeon and an additional ship's mate. These two groups of slaving mariners were those most likely to collect the natural historical specimens that eventually joined British museums and gardens. Understanding the shifting geographies of the slave trade over the course of the eighteenth century can explain the provenance of specimens acquired during this period. The evolving rhythms of the British slave trade illuminate why specimens were gathered in particular places along the West and West Central African coast. Further, the priorities of British slaving companies and imperial officials tended to encourage some areas of natural study over others. Distinctive aspects of the transatlantic slave trade shaped the collections and knowledges that British naturalists obtained through its circuits.

Captivity's Collections is located at the intersection of the history of science and the history of the slave trade. Until relatively recently, the rich and dynamic scholarship associated with the histories of science and the slave trade developed largely separate from one another. There are multiple reasons for this, including the typical training scholars in the two fields once received and the organization of relevant archival materials. The history of the history of science and of abolitionism provides an additional perspective. The first historians of science were practicing natural philosophers who, beginning in the late eighteenth century, used the history of science as a way to validate their intellectual pursuits at a time when the study of the natural world was not universally accepted as a worthy endeavor. Early historians of science represented the study of the natural world as a progressive force that benefited and improved society. They proclaimed that natural philosophy and natural history would be responsible for ameliorating the human condition and framed the history of the field as a story of "steady upward progress."[17] Around the same time, abolitionists and antislavery advocates in Britain and the United States marshaled scientific arguments in support of the abolitionist cause and depicted slavery as premodern, the enemy of progress, and "at odds with a world defined by science and technology." Modern scholarship has repeatedly and definitively shown such a depiction of slavery to be a myth not rooted in historical fact. Yet the idea of slavery as fundamentally at odds with science has had a long half-life and continued to influence both scholarly and popular ideas about slavery into the twentieth century.[18]

Transformative recent work in slavery studies argues for the fundamental importance of slavery and the slave trade to the development of modern economic, commercial, social, and political institutions around the Atlantic basin. It demands that we reject the idea of slavery as a "peculiar institution." In other words, it requires that we abandon the notion that slavery was anomalous or incidental to the development of the modern world. Instead, it challenges us to wrestle with the notion that slavery and the slave trade were among the foundations upon which the modern world was built. This work has demonstrated the ways in which capitalism as well as ideas about credit, insurance, labor, free trade, individualism, and freedom developed out of the commerce in captive Africans.[19] *Captivity's Collections* builds on this scholarship by highlighting the role played by slavery in the development of scientific institutions. It demonstrates how the British slave trade was instrumental to the production of natural knowledge and to the acquisition of natural historical specimens and observations that became part of scientific museums, botanic gardens, herbaria, and texts.

Transformative recent scholarship on slavery's legacies has also sought to understand the economic and cultural impact in Britain of the slave economy. This work has painstakingly traced the economic, political, cultural, and philanthropic legacies of wealth derived from British slavery. The Legacies of British Slave-ownership project at University College London, for example, revealed the influence of wealth produced by British slavery using the records of the Slave Compensation Commission.[20] The commission was charged with distributing £20 million in compensation for British enslavers, which was appropriated by Parliament in 1833 when it abolished slavery in

British colonies. The publicly accessible database that formed the foundation of the Legacies of British Slave-ownership project records the 47,000 individuals who made a claim for compensation and provides biographical details for 3,000 of these individuals. This data enabled analyses of the impact of wealth derived from slavery on British culture and society. It links significant writers, politicians, historians, imperial administrators, and commercial and financial firms of Victorian (and modern) Britain to slavery.[21] Similarly, in September 2020 the National Trust released an assessment of the historical ties between its ninety-three historic properties and British slavery and colonialism. Scholarship on British slavery's legacies has documented ties between the slave economy and British country houses, art collections, political careers, and generational wealth. We know much less about slavery's legacies for British science and scientific institutions.[22]

Examining the entangled histories of British science and the slave trade opens up the possibility of seeing old questions in the historiography of slavery in new ways. For nearly eighty years, historians, economists, and other scholars have debated the profitability of the slave trade and its role in Britain's economic development and industrialization. This debate dates to Eric Williams's pioneering *Capitalism and Slavery*, published in 1944. Williams argued for the central importance of capital accumulated from the slave trade and plantation slavery in the development of the Industrial Revolution and more generally in reshaping the British economy.[23] *Captivity's Collections* highlights the need to think broadly about the profits of the slave trade. Collections such as that gathered by Burnet on the *Wiltshire* demonstrate that the proceeds of the slave trade that accrued to Europe and its American colonies through the exploitation of Africa and its peoples transcended economics. To fully understand how individuals and institutions on both sides of the Atlantic profited by the slave trade, we must include collections of rare specimens and the scientific knowledge that resulted from their study among slaving's profits.

Williams argued that the wealth of individual enslavers reshaped British society through its investment in country estates, political careers, philanthropy, and patronage of the arts. These claims have been substantiated in stunning detail for the early nineteenth century by the Legacies of British Slave-ownership project. Other scholars, including James McClellan and James Delbourgo, have shown how wealth derived from the slave economy funded the pursuit of natural history and natural philosophy in the early modern period.[24] *Captivity's Collections* makes a complementary but distinct argument. It contends that natural historical collections themselves represent a form of profit derived from the slave trade. The capital that accumulated in Britain and British North America because of the slave trade included natural knowledge and specimens used to produce that knowledge. Williams illuminated how the histories of slavery and capitalism intertwined in the eighteenth century to help create the modern world. *Captivity's Collections* argues that science, another deeply transformative force in the creation of the modern world, was inextricably linked with the slave trade.

Captivity's Collections joins the growing scholarly literature that reveals how early modern science was built upon slavery and the slave trade. Literary scholars and art historians have shown how the violence upon which slavery rested was inscribed in the texts and images produced by naturalists working in slave societies.[25] Historians of medicine have revealed the slave trade as a transformative site of medical and pharmaceutical knowledge production. They have traced, for example, how deadly disease environments along the routes of the slave trade reshaped medical frameworks, upended older ideas about the relationship between environment and disease, contributed to a medical construction of race, encouraged a cosmopolitan and promiscuous approach to healing techniques, and dramatically reshaped the global trade in medical substances.[26] Other scholars have revealed ways that enslaved collectors, informants, and healers played crucial roles within networks of natural history and in the development of geographical knowledge. Naturalists in plantation societies appropriated natural and medical knowledge from enslaved Africans as well as depended upon seeds and specimens gathered by enslaved collectors.[27] Yet much of this work, especially that on natural historical collecting, has focused on plantation slavery rather than the transatlantic slave trade itself. *Captivity's Collections*, by contrast, reveals how naturalists exploited the networks of the British slave trade to acquire seeds and specimens for their museums, gardens, and herbaria.

The circuits of the British transatlantic slave trade crossed political and imperial boundaries. The inhuman trade in African captives brought British slaving mariners to portions of the Atlantic World claimed by rival imperial powers as well as to regions where Europeans effectively projected little or no imperial power. *Captivity's Collections* follows the geographies of the British slave trade rather than being constrained by modern political boundaries or by early modern imperial claims. It reflects the call within the historiography of the Atlantic World to center the Atlantic Ocean itself. By so doing, it highlights the porousness of imperial boundaries, the malleability of identities, and the importance of smuggling in the early modern Atlantic. It also helps to bring into focus the role of mariners within the circuits of early modern science.

Slaving mariners like Burnet fit uneasily into many of the spatial frameworks common within scholarship on science and empire. An older tradition within the history of science focused on questions of how scientific knowledge and practice diffused from European centers to colonial peripheries. It tended to view scientific knowledge as stable, universal, and produced in Europe. More recent scholarship understands all scientific knowledge as locally produced. It asks how knowledges move from one locality to another and how one local knowledge comes to be understood or claimed as universal. Recent scholarship on science and empire highlights the polycentric networks of natural inquiry and the transformation of local knowledges as they circulated through those networks.[28] Much of this scholarship takes one or two terrestrial spaces as the locality under study. As professional itinerants, slaving mariners regularly moved between spaces and empires, although the time they spent in any one of them could be relatively short. Burnet, for example, collected in Anomabu, São Tomé, Buenos Aires, the British Caribbean, and from the Atlantic Ocean itself in the span of just two years. Slaving mariners also regularly moved between imperial centers and peripheries, and thus belonged fully to neither space. *Captivity's Collections* builds on Kapil Raj's observation that circulation can itself be a site of knowledge production by focusing on the British transatlantic slave trade as a space of natural history.[29]

― ― ― ― ― ― ― ― ― ― ―

We do not often think about the wretched, miserable, and inhuman spaces of slave ships as simultaneously sites of natural history. Collections such as that gathered by John Burnet on board the *Wiltshire* suggest that this is exactly what they were. The pages that follow document how thousands of specimens were obtained by slave traders, slave ship captains, slaving surgeons, and other individuals engaged in the British transatlantic slave trade. The collecting practices of individuals along the routes of the slave trade often included the appropriation of African knowledges and relied on the labor and expertise of individuals unacknowledged in surviving historical sources. Natural historical specimens, along with the written descriptions, observations, and letters that often accompanied them, were transported on the same vessels on which captive Africans who had been ripped from their families and homes endured the inhuman crowding, death, disease, and violence that characterized the transatlantic slave trade. *Captivity's Collections* also reveals how directors of British slaving companies and metropolitan naturalists sought to use natural history to increase the profitability of the slave trade. Company directors encouraged slave traders to collect specimens and natural knowledge in order to identify drugs, dyes, and other natural commodities in which they might trade.

The collection that Burnet acquired on board the *Wiltshire* is simultaneously unique and representative of the other collections chronicled in these pages. Like many who collected specimens along the routes of the British transatlantic slave trade, Burnet was a slave ship surgeon. Slave ship surgeons were typically the only members of a slaving crew with some training—however rudimentary—relating to natural history. Much of the value of the *Wiltshire* collection came from the inclusion of specimens from West Africa and Spanish America, regions from where British naturalists had received relatively few natural historical specimens. Similarly, the value of most collections gathered by British slave traders lay in their rarity. The inhuman commerce in captive Africans facilitated the movement of British slaving mariners throughout the Atlantic World and gave them access to places where Britons were otherwise unlikely or unwelcome to visit, such as Anomabu and Buenos Aires. While most slaving mariners did not collect specimens of the flora and fauna they encountered on such voyages, the few who did collect had an outsized impact on the development of natural history. Further, naturalists were not the only ones interested in the biota loaded into the holds of slave ships. British slaving companies were keenly interested in specimens that had the potential to become profitable natural commodities, like the scarlet red dye in the *Wiltshire* collection.

Yet the *Wiltshire* collection is also unique among the examples discussed in this book in two important ways that bear consideration. First, while most accounts of collecting slave traders are fragmentary, the story of the *Wiltshire* collection is relatively complete. We know with an unparalleled level of detail what was collected, where, by whom, and what happened to it upon reaching Britain. Further, our knowledge of John Burnet's personal history is quite rich compared to what we know about most slaving mariners who collected specimens. This is largely due to unique aspects of Burnet's long and eventful career.[30] Taken together, these sources make it possible to reconstruct the *Wiltshire*'s voyage and the collection that resulted from it. By contrast, the records relating to the slave ship surgeon John Kirckwood are more representative of the sources underpinning this book. Like Burnet, Kirckwood worked as a British slave ship surgeon and obtained specimens on behalf of James Petiver in the early eighteenth century. Yet references to Kirckwood's collecting activities are few, fragmentary, and frustratingly brief. The longest reference to Kirckwood is found in Petiver's 1702 publication, *Gazophylacii Naturae*. In it Petiver reports that he received an Angolan butterfly from the slave ship surgeon. He describes the insect as the "Papilio ANGLOENSIS, margine pulchre oculate," and notes "Mr. John Kirckwood Surgeon was the first who discovered this elegant fly at Angola."[31] Surviving sources tell us little about Kirckwood besides the fact that he was employed as a surgeon and traveled to ports along the routes of the British slave trade.

More importantly, the *Wiltshire* collection is unique among the examples discussed in this book because there is clear evidence that the collection included human specimens. The collection Burnet gave to Petiver included an aborted fetus and three polyps removed from the hands of captive Africans. All four of these specimens likely came from Africans on board the *Wiltshire*. Human remains and other surgical specimens were commonly found among the collections belonging to early modern medical professionals. Physicians and naturalists working in New World slave societies sometimes included in their museums specimens that had been taken from the bodies of enslaved Africans. Such surgical and anatomical specimens reflected and reinforced the inequalities of power and the exploitation of Black bodies fundamental to the chattel slave system.[32] It would stand to reason that British collectors who engaged in the slave trade, especially those with medical training, also acquired specimens extracted from the bodies of captive Africans even if evidence of these collecting practices is sparse.[33]

Specimens such as the fetus and polyps in the *Wiltshire* collection make tangibly clear that the slave trade's violence and exploitation enabled the production of new natural knowledge. The fact that most of the specimens discussed in the chapters that follow are butterflies, seeds, shells, and plants should not distract from the context of their collection and transportation. As Pratik Chakrabarti has argued, the theme of violence and exploitation is an important one within the history of science, albeit one infrequently acknowledged.[34] The violence, coercion, torture, and death that characterized the *Wiltshire* also characterized the hundreds of other slave ships on which specimens and the individuals who obtained them traversed the Atlantic basin.

Fundamentally, this is a book about how British naturalists exploited the commercial networks of the transatlantic slave trade to obtain natural historical specimens and thereby to produce new natural knowledge. It is primarily engaged in recovering the history of how British natural historical collecting in the eighteenth century relied upon the inhuman commerce of the slave trade. It traces how natural historical specimens became part of British collections and gardens. Consequently, it focuses primarily on British slaving mariners, slaving agents, naturalists, and others relying on the routes of the slave trade to obtain natural historical specimens. Collecting slave traders relied upon individuals of African descent as informants, guides, and knowers. These often anonymous contributions are acknowledged throughout *Captivity's Collections* but are not the primary focus of the book. Similarly, it is not intended as a contribution to the rich literature of middle passage studies that seeks to understand "slavery at sea," nor does it focus on the lived experiences of enslaved Africans whose brutal exploitation enabled natural historical collecting.[35]

Instead, *Captivity's Collections* shows how the transatlantic slave trade facilitated the seemingly distinct efforts of British naturalists to acquire new natural historical specimens and to produce new natural knowledge. It demonstrates how quotidian aspects of the British slave trade, such as its routes, ports, and personnel, shaped the collections made by slave traders. The transatlantic slave trade served as conduit, bringing West African, Spanish American, and British American plants, animals, and shells into European collections. The book's first chapter examines how British naturalists at the turn of the eighteenth century exploited the routes of the British slave trade to obtain natural historical specimens from ports in West and West Central Africa. Metropolitan naturalists in the late seventeenth century largely relied upon employees of the Royal African Company to collect on their behalf in Africa. As the British slave trade expanded in the early eighteenth century, so too did naturalists' sources for African specimens. Naturalists increasingly relied upon slaving mariners on board ships owned by independent slave traders to acquire floral, faunal, and mineral specimens. West and West Central African specimens gathered by slave traders in the early eighteenth century were springboards for discussion in scientific societies, appeared in scientific texts, and became part of natural historical museums and botanical gardens.

The Royal African Company lost its monopoly on the legal English trade to West Africa in 1712. A handful of the company's investors believed that natural history would enable the company to find new ways to profit from its trade. Chapter 2 examines how the company's leaders attempted to employ the practices of natural history to usher in a new period of profitability. They instructed the company's slaving agents to search for botanical and mineral resources indigenous to West Africa and encouraged the cultivation of plantation products at the company's West African trading factories. Their plans relied upon the appropriation of African knowledge, the commercial infrastructure of the slave trade, and the collection of natural objects.

The next two chapters focus on how British naturalists exploited the *asiento* agreement, which gave the British South Sea Company a monopoly on the sanctioned slave trade to Spanish America between 1713 and 1739. Chapter 3 traces the efforts of company employees to leverage their unusual access to Spanish America into new scientific knowledge by surreptitiously collecting flora and fauna. Although only a handful of South Sea Company employees undertook such investigations, their efforts uniquely shaped British natural history. The collecting activities of company employees might have been illicit, but their presence in Spanish America was aboveboard. In contrast, chapter 4 follows the collecting efforts of William Houstoun, a botanist and former South Sea Company surgeon who returned to Spanish America under the cover of the *asiento* slave trade. A group of British patrons underwrote Houstoun's efforts to collect illicitly "all the useful plants" in Spanish America and to introduce them into British gardens and the new colony of Georgia. Houstoun and his successor Robert Millar sought to smuggle some of the most valuable natural commodities known to the early modern world, including cinchona, cochineal, ipecacuanha, and jalap, by means of the British slave trade to Spanish America.

One of the largest known beetles was found floating in the Gabon Estuary in 1766 by a British ship captain. The Goliath beetle quickly became an object of desire among natural historical collectors. Chapter 5 traces the efforts of Dru Drury, a British silversmith and entomologist, to acquire a specimen of the Goliath beetle for himself. The silversmith's correspondence, account books, and museum inventory allow us to trace how a collector in the mid-eighteenth century might utilize British commercial and naval circuits to Africa in the pursuit of a particular specimen. The dramatic expansion in British participation in the slave trade by the middle of the century meant that naturalists such as Drury found new ways to exploit British commercial networks in order to collect specimens from a distance.

Drury ultimately concluded that the best way to obtain rare African specimens like the Goliath beetle was to send a British naturalist to the continent. Consequently, Drury and a small group of patrons sponsored Henry Smeathman's four-year collecting expedition to Sierra Leone from 1771 to 1775. Chapter 6 relies upon Smeathman's journal and correspondence to reveal how a naturalist funded by patronage was still dependent on the commercial infrastructure of the slave trade in order to collect, classify, sketch, preserve, and study rare insects and other specimens.

As these accounts illustrate, thousands of natural historical specimens were obtained through the British transatlantic slave trade and added to European museums, botanic gardens, and herbaria during the eighteenth century. Some of these specimens survive into the present day in the collections of institutions such as London's Natural History Museum, the Hunterian Museum at the University of Glasgow, and the Oxford University Herbaria. Most modern scientific researchers who consult these collections, like the visitors who flock each year to the scientific museums that own them, are unaware of how institutions acquired these specimens in the first place. Pressed plants, seeds, shells, insects, and other natural objects obtained by means of the slave trade sit as largely unacknowledged monuments to the exploitation of millions of Africans and to the immense sweep of the inhuman commerce that brought natural historical specimens into captivity's collections.

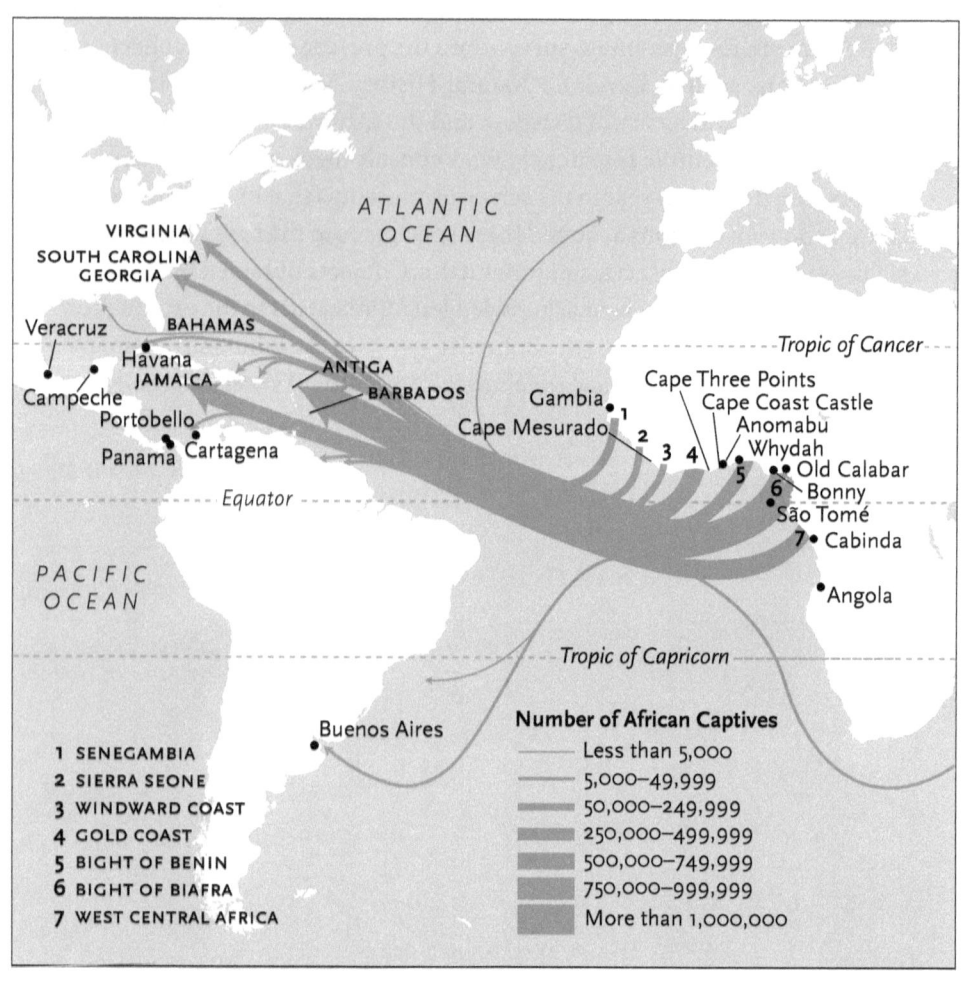

MAP 1.1. Major routes and ports of the British transatlantic slave trade, ca. 1563–1810, and provenance of natural historical specimens. Place-names indicate locations where specimens in British collections were obtained by slaving mariners and agents.

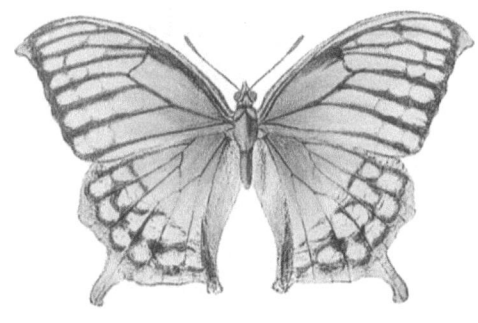

1.

Cannot on These Coasts Gather Amiss

In the fall of 1710 Capt. Thomas Johnson supervised his crew as they prepared the *New Providence* for its voyage to West Africa (map 1.1). The *New Providence* was headed to Gambia, where Johnson purchased 225 captive Africans. Four and a half months after leaving England, Johnson oversaw the disembarkation in Kingston, Jamaica, of the 180 enslaved Africans who had survived the middle passage. During the months the *New Providence* spent along the African coast and crossing the Atlantic, responsibility for keeping both captives and crew alive fell to Robert Barcklay, the ship's surgeon. A slave ship surgeon like Barcklay would have spent his time assisting the captain in the purchase of enslaved Africans, inspecting captives on board ship for sickness, administering medicine (often against captives' wills), and overseeing their feeding and force-feeding. Such were Barcklay's obligations to the slave ship captain.[1]

Barcklay, however, was under a second obligation. He had promised apothecary and naturalist James Petiver that he would obtain natural historical specimens during the voyage. While the *New Providence* prepared to sail in 1710, Petiver reminded Barcklay of his pledge. The naturalist sent Barcklay "some few Tracts relating to Natural History" as well as images of "Shells

Insects Plants &c as are to be met with in the parts you are going to." Petiver expected that such publications would guide Barcklay's collection of West African specimens. But he assured the ship surgeon that whatever he acquired in Gambia, "tho never so common," would be greatly appreciated as he had "yet never seen anything from that part of Africa." Once Barcklay reached Jamaica, Petiver hoped that he would help him collect more collectors. He urged Barcklay to present the texts he sent to any "Curious" person he encountered who agreed to also collect on his behalf. The naturalist pledged to publish images of specimens Barcklay gathered during what he "heartily pray[ed]" would be "a prosperous Voyage."[2]

A profitable slaving voyage for Petiver was not measured in pounds and pence, nor in the number of captive Africans forcibly transported to Jamaica. Rather, it was calculated in rare natural historical specimens added to his museum and new correspondents who might themselves acquire flora and fauna on his behalf one day. Petiver relied on the British slave trade to obtain natural historical specimens from Africa and the Americas for nearly thirty years.[3] British slaving mariners, surgeons, and agents acquired specimens, appropriated knowledges, and recruited new collectors on Petiver's behalf. They also played crucial roles as intermediaries within the naturalist's broader network of collectors in the Atlantic World.

Petiver was not alone in his dependence upon the routes and individuals of the British slave trade at the turn of the eighteenth century. Contemporaries and fellow naturalists including Sir Hans Sloane, Dr. James Douglas, William Sherard, and Leonard Plukenet similarly obtained specimens from individuals engaged in the transatlantic slave trade. Many of these naturalists were members of the Royal Society of London, the leading scientific society of the English-speaking world. The society's meeting minutes testify to the regularity with which specimens and observations obtained along the routes of the British slave trade became the subject of fellows' discussions. Metropolitan naturalists assured slaving mariners that they could not go wrong gathering flora and fauna in West and West Central Africa.[4] As Petiver told Barcklay, even Gambia's most common plants would be very welcome because metropolitan naturalists knew so little about Africa's natural productions. Petiver hoped that the slave ship surgeon would acquire many of the valuable plants depicted in the natural historical texts he provided. But Petiver also understood, as he told another slave ship surgeon, that "the rest will be very acceptable since you cannot on these coasts gather amiss."[5]

Petiver did not necessarily expect that slaving mariners like Barcklay would personally gather African flora and fauna. Expropriated knowledge and labor were frequently part of the material practices of natural historical collecting in the Atlantic World. This was especially true along the routes of the slave trade. Free and enslaved Africans identified, obtained, and preserved many of the natural historical specimens from West and West Central Africa that slaving mariners and agents like Barcklay later sold or gave to European naturalists. Decisions about which specimens should be gathered and where best to search for them were often based on African natural and medical knowledges. Although African collectors were almost always anonymous and frequently invisible in surviving historical sources, their knowledge, expertise, and skill were materially embodied in the specimens they collected. The floral, faunal, and mineral specimens they gathered represented another form of profit extracted from the holds of slave ships.[6]

Metropolitan naturalists such as Petiver understood that slave ships like the *New Providence* were also potential sites of natural history. Distinctive aspects of the British transatlantic slave trade to West and West Central Africa shaped the collections and knowledges produced along its routes at the turn of the eighteenth century. In the late seventeenth century, metropolitan naturalists largely relied upon employees of the Royal African Company, which held the monopoly on the (legal) British trade with West Africa. The Royal African Company lost its monopoly in 1698, which led to an astounding growth in the British slave trade. Metropolitan naturalists increasingly turned to slaving mariners employed by independent traders to acquire specimens and to recruit new collectors in the early eighteenth century. The expansion of British slaving during the early eighteenth century meant more warfare, violence, and human suffering for people living in places like the Gold Coast. For British naturalists, increasing numbers of British vessels and mariners engaged in the transatlantic slave trade meant more opportunities to exploit these commercial routes in order to acquire rare flora and fauna. The geographies and dynamics of British slaving and natural historical collecting shifted as free trade replaced the Royal African Company's monopoly over the course of the early eighteenth century. The understanding that natural historical collecting could be integrated into the business of enslaving remained constant.

COLLECTING AT CAPE COAST

In 1693 the Reverend Dr. John Smyth had nearly completed his posting as the Royal African Company's chaplain at Cape Coast Castle, a British slaving fort on the Gold Coast and the Royal African Company's headquarters in Africa.

Before he departed Cape Coast, Smyth fulfilled a promise he had made to Petiver: to acquire West African plants on the naturalist's behalf. The minister obtained forty-six plants that were esteemed by Africans along the Gold Coast for their medicinal properties. Each specimen was pressed to preserve it and had a label affixed that indicated its local name, its medicinal properties, and the recommended method of preparing the medicament derived from it. Smyth shipped the entire collection of West African plants that had been so carefully collected, preserved, and labeled to Petiver in London. It is clear that Smyth's collection was based on the medical and natural knowledge of Africans living along the Gold Coast, probably the Fante people. Less clear is who actually identified, gathered, and preserved the herbaria specimens. While Smyth himself may have done so, it is more likely that he employed a free or enslaved African to undertake some or all of this physical and intellectual labor. The herbaria specimens sent by Smyth, with their neat labels recording African medical knowledge, exemplified in material form the appropriation of African plants and African natural knowledge.[7]

Collections such as that sent by Smyth enriched the gardens, herbariums, museums, and publications upon which the work of natural history depended. Such collections also reflected the importance of Royal African Company employees within late seventeenth-century English networks of natural history. During the early 1690s almost all of Petiver's African specimens came from Royal African Company employees stationed at Cape Coast Castle. Most West African specimens presented to or discussed before the Royal Society in the late seventeenth century can similarly be traced back to employees of the Royal African Company.

The Royal African Company dated to the late seventeenth century, although the origins of English participation in the transatlantic slave trade were much older. The first English slave-trading voyages were those commanded by John Hawkins in the 1560s. Over the next century, English commerce with West Africa was intermittent, small in scale, and often not focused on slaving. Sustained English trade with West Africa began with the creation of the Company of Royal Adventurers Trading to Africa in 1660. The company's royal charter gave it a monopoly on English trade in West Africa and was intended to encourage more regular commerce in gold and, to a lesser extent, enslaved Africans. The company counted the king's brother, James, Duke of York; his cousin, Prince Rupert; and numerous courtiers among its investors and directors. Despite these well-placed supporters, war with the Dutch and insecure finances led to the reorganization of the company as the Royal African Company in 1672. The new company's royal charter similarly gave it a monopoly on all English trade with West Africa. Like the various commercial ventures that preceded it, the Royal African Company traded in natural commodities such as gold, dyewoods, and ivory. Unlike previous English ventures, the commerce in enslaved Africans was its focus.[8]

The Royal African Company in the 1690s sought to profit from the sale of human beings and the trade in gold, not to wield imperial power. European presence in West Africa during the seventeenth and eighteenth centuries was largely confined to a series of trading forts or factories along the coast. These factories are better understood as fortified warehouses than as military strongholds that could command obedience over the surrounding countryside. European trading companies paid ground rent, customs duties, and other forms of tribute to local African rulers in exchange for the right to remain at their factories. They also constantly had to contend with one another, especially along the Gold Coast. Although the Gold Coast is less than 300 miles long, European trading companies from five different nations built and attempted to maintain almost sixty forts along its length during the seventeenth century. The largest, best fortified, and most important among the English forts in West Africa was Cape Coast Castle (fig. 1.1). Its halls served as home for the dozens of white and Black traders, artisans, linguists, accountants, clerks, soldiers, and domestic servants employed by the company, while its dungeons confined hundreds of captive Africans bound for the New World.[9]

Coastal fortresses like Cape Coast Castle have come to symbolize the violence, horror, and expropriation of the transatlantic slave trade. Yet most were originally built for the trade in gold, not captives. For nearly two centuries Europeans traded along the Gold Coast primarily in the precious metal that inspired the European name for the region. The Gold Coast in the eighteenth century came to be dominated by the Fante people who spoke one of the Akan languages, although there is evidence of additional linguistic and cultural diversity in the region. Akan gold traders were known for their acumen as traders and for the precision of the scales they used to weigh gold. Despite the long history of the Afro-European gold trade, the transition to human trafficking as the dominant transatlantic trade in the region was abrupt, dramatic, and violent. While the eighteenth century generally saw an expansion of slave trading in West and West Central Africa, "nowhere else did the volume increase as much and as quickly as it did on the Gold Coast." In most regions of West and West Central Africa the scale of slave trading increased over multiple decades; in the Gold Coast, the transition occurred largely during one decade, the 1690s. The Fante's long experience trading

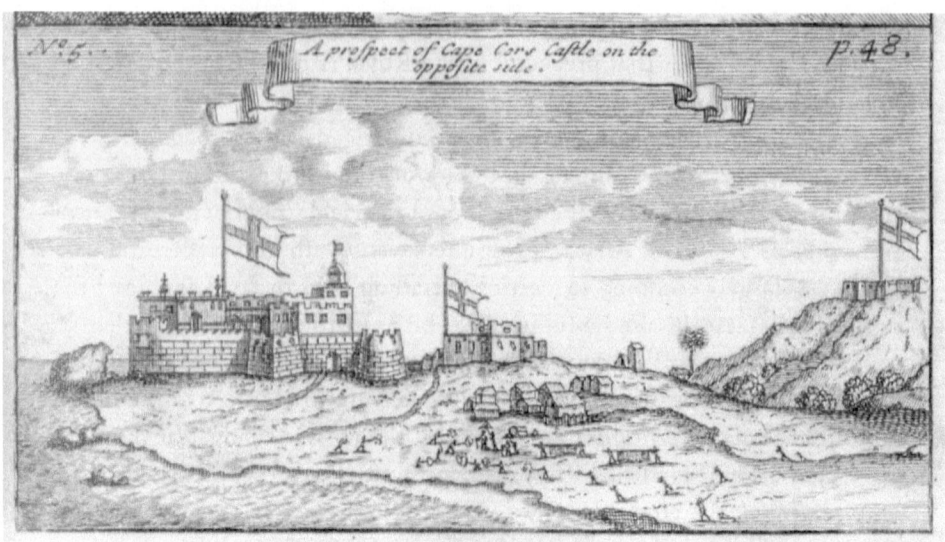

FIGURE 1.1. Cape Coast Castle. Willem Bosman's image of Cape Coast Castle (*left*), the African headquarters of the Royal African Company, also prominently featured Edward Bartar's fortified home (*center*). During the 1690s, Bartar both obtained plants, shells, and insects for British naturalists and was a powerful figure within the slave trade on the Gold Coast. Detail from Willem Bosman, *A New and Accurate DESCRIPTION of the Coast of Guinea*, London, 1705, RB 67988, Huntington Library, San Marino, California.

with Europeans helps to explain the distinctive features of the political order that developed as a result of the transition to large-scale involvement in the transatlantic slave trade. It also meant that the Fante who English natural historical collectors like Smyth relied upon were Atlantic creoles, whose linguistic, cultural, and commercial skills would have been as valued by naturalists as they were by European merchants.[10]

Royal African Company slaving agents, ministers, and surgeons were some of the Englishmen who spent the most time in Africa in the late seventeenth century. Consequently, they were important sources of information about the region's peoples, geography, and natural world. The geographic scope of the Royal African Company's commercial activities meant that most of the African collections and observations received by the Royal Society in this period concerned the Gold Coast and, to a lesser extent, Gambia. The Royal Society's museum included the skin of a large African snake presented by Capt. John Smith and a spiral-shaped elephant's tusk given to the society by Thomas Crispe, a member of the Royal African Company's governing body. In the early 1670s the Royal Society's secretary, Henry Oldenburg, sent the African Company's minister at Cape Coast a series of questions about West Africa and the people who lived there. Oldenburg's "Inquiries Concerning Guiny" included questions about the geography, trade routes, climate, minerals, and animals of the West African coast as well as about the "Genius, Customs, [and] Exercises" of the local peoples.[11] A decade later John Flamseed, the astronomer royal, shared with the Royal Society astronomical observations he had received from a Mr. Heathcote, employed by the Royal African Company at Cape Coast. A Mr. J. Hillyer, also stationed at Cape Coast, sent a seven-page weather journal. Another company employee, Agent Greenhill, reported to the society that he discovered a large mound of oyster shells several miles inland "which being burnt made him a most Excellent sort of Lime."[12] The Royal Society turned to Royal African Company employees for information about a wide range of topics relating to West Africa's natural world.

James Petiver was foremost among the British metropolitan naturalists who turned to Royal African Company employees as collectors in the late seventeenth century. Extracting promises to obtain natural historical specimens from merchants, mariners, colonists, and other travelers lay at the heart of Petiver's collecting strategy. By the time of his death in 1718, Petiver acquired one of the largest and most diverse natural history collections then known in Britain. Twin-forked stag beetles from the East Indies, shells from the Caribbean, ferns from Cochinchina, and moths from the English countryside appeared side by side in his museum and in the illustrated catalogs of it that he frequently published. His herbarium alone is estimated to contain over 21,000 specimens. In his publications, Petiver frequently repeated his request that "*Practitioners* in *Physick, Sea-Surgeons,* or other *Curious Persons* who Travel into *Foreign Countries,*" would "be pleased to make Collections for me of whatever *Plants, Shells, Insects* &c. they shall meet with."[13] The size and scope of Petiver's museum might suggest a collector of great wealth or social connections. Yet the apothecary was a man of relatively modest means. What he lacked in social standing and education, he made up for in the tenacity with which he pursued new natural curiosities and those who could provide them. The secret to Petiver's success as a collector lay in his dogged enlistment of individuals who plied the global routes of British commerce, including the slave trade.[14]

Chief among the Royal African Company employees Petiver recruited to collect for him was Edward Bartar. The many years Bartar lived at Cape Coast Castle made him a particularly valuable addition to Petiver's network of collectors and correspondents. Only a handful of Britons who traveled to West Africa in the late seventeenth century measured their time on the coast in years rather than months. Bartar, however, was unlike most of Petiver's correspondents. He was of Anglo-African descent, probably born to an English slave-trading father and an African mother in West Africa. As a young man Bartar traveled to England to be educated at the expense of the Royal African Company. During this time, he seemed to have moved in some of the same botanically inclined social circles as Petiver. Bartar returned to the Gold Coast in 1693 as an employee of the Royal African Company drawing an annual salary of thirty pounds.[15] The company's records reveal Bartar to be an agent tasked with a wide range of responsibilities. He was entrusted to collect debts owed by local African and European traders, to inventory trade goods, to purchase corn and cattle, to transport enslaved Africans within the Gold Coast, to negotiate peace between the company and local Africans, and, when that failed, to help defend its factories against attack.[16] In time Bartar came to be more powerful than his nominal bosses, the company's agents at Cape Coast Castle. According to one contemporary, "whoever designs to Trade with the *English* must stand well with him before he can succeed." The extent of Bartar's influence came even to be inscribed on the built environment. Just below Cape Coast Castle in the engraving of the factory shown in figure 1.1 lies a second fortified structure that was in many ways a smaller version of the company's fort, complete with an English flag and canons. This miniature version of the castle was Bartar's home, where at the height of his influence he commanded a small personal army and oversaw much of the local slave trade.[17]

Petiver always expected great things from those who pledged to collect on his behalf, but his hopes for Bartar were even grander than usual. The naturalist's optimism about Bartar's potential as a collector seems to have been based not on the influence Bartar commanded along the Gold Coast but on the London social circles in which he had once moved. During the years Bartar had lived in England, he had befriended avid naturalists and gardeners whom Petiver also counted among his close friends. The apothecary tried to motivate Bartar to collect more frequently by reminding him of their many friends and acquaintances eagerly awaiting each new package from Cape Coast. For example, just months after Bartar's return to Africa, Petiver declared that apothecary Samuel Doody, professor of botany Leonard Plukenet, "and several other friends . . . once a week remember you in a Glass of Nottingham Ale" and looked forward to receiving his collections.[18]

A few years later, Petiver and his friends held out the possibility of membership in the Royal Society of London as an inducement to collect. The African Company chaplain, Smyth, reported from London that several of the society's fellows "design to make you one of them which will be a means most highly to advance you here." Smyth recommended that the slaving agent "leave no stone unturned to serve them" by gathering a large collection. The former Cape Coast minister recommended Bartar send snakes, garlic bark, ferns, fruit, and other "herbs," "the grass that negroes wear at their sides which smells like new hay," and "many things that you have there you'll not think worthy your interest." Smyth prophesied that one large collection sent to the Royal Society would "redound more to your credit & honor" than all Bartar could hope to earn in the Royal African Company's service.[19]

When the reality of Bartar's collecting did not live up to Petiver's expectations, the naturalist's disappointment was all the greater. "I cannot so much blame a Stranger who knows nothing of the way of gathering Plants" and whose stay in West Africa was very short. However, the naturalist continued, when "you that know so well and have so many frequent opportunities of making me Collections not only of Plants but Shells & Insects, have ... forgott your friend," Petiver's reproach was unmistakable. Adding to the apothecary's frustration was that some of the specimens that he did receive from Bartar were incomplete, consisting of "only leaves and wanting either flower or fruit or both." As he petulantly reminded Bartar, "you very well know" that he desired natural historical specimens "beyond all your Gold in Africa."[20]

From Petiver's perspective, Bartar was perfectly positioned to become a significant collector. The importance of Cape Coast Castle within the networks of the slave trade meant that ships frequently called at the slaving fort, including those bound for ports outside of the English empire. Petiver hoped that Bartar would recruit ship captains sailing to Brazil, São Tomé, Angola, and other "parts remote or far distant" to collect specimens on his behalf. The London naturalist also expected that Bartar would gather a large collection of West African flora and fauna.[21]

Petiver acknowledged that Bartar's duties for the Royal African Company consumed most of his time, so he urged Bartar to have one of his enslaved people collect in his stead. He recommended that each month Bartar send an enslaved African into the interior, equipped with a basket and three or four quires of brown paper. Enslaved collectors, the apothecary recommended, should gather all the ferns, mosses, trees, shrubs, and herbs they could find, pressing plants' leaves and flowers between sheets of brown paper and filling the basket with the plants' fruits.[22] When Petiver started to doubt that Bartar was following his advice, he switched tactics. Slave ship surgeons who called at Cape Coast Castle reported that Bartar was continually attended by three or four enslaved Africans. Why, asked the London apothecary, could these individuals not assist visiting slave ship surgeons by collecting on behalf of Petiver? Instead, Petiver fumed, Bartar hindered the surgeons "from making collections . . . by your promising you would do it yourself & denying them your assistance of your own slaves which you could so well have spared to doe."[23] Petiver's comments reflected the frequency with which naturalists in the Atlantic World employed enslaved Africans as collectors, hunters, guides, and technicians who prepared specimens for transport. Metropolitan naturalists exploited the commercial infrastructure of the transatlantic slave trade as a means to obtain specimens from West Africa and from throughout the Atlantic World. This included actively and directly exploiting the labor of enslaved peoples.[24]

Contrary to what Petiver's frequent complaints might suggest, Bartar kept his promise to collect specimens on behalf of his friends, albeit not at the level they had hoped. Within two years of the slaving agent's return to the Gold Coast, Petiver published descriptions of the trefoil ground-bean, the matice-weed, the scorpion senna, and the Malabar bindweed plants based on specimens Bartar had gathered in the vicinity of Cape Coast Castle.[25] Petiver shared the specimens he received with other members of the Royal Society. Writing from London in 1696, the former Royal African Company minister Smyth reported that members of the Royal Society believed Bartar "had honored them highly in sending Mr. Petiver so brave a collection of plants." Petiver assured Bartar that his collection of plants from Cape Coast included many that were curious and strangers to European botany, high praise indeed coming from Petiver.[26] Over the next few years Bartar continued to send his English friends plants, shells, butterflies, and other insects. These collections included "an elegant hairy Catterpillar," "Bartars dark Guinea Butterfly with white spotts," and the Pintado butterfly. It also included three or four quires of pressed plants valued by the peoples living in the vicinity of Cape Coast Castle for their medicinal virtues.[27]

Just over a dozen of these specimens can today be found in the Sloane Herbarium at the Natural History Museum in London. Among them are trefoil ground-bean, the matice-weed, and two leaves labeled in an unknown hand: "The juce of this leaf is good for sore eyes the name unknown" (fig. 1.2). The label's author (very possibly Bartar) highlighted the healing properties of the leaves based on how the plant was used along the Gold Coast, most

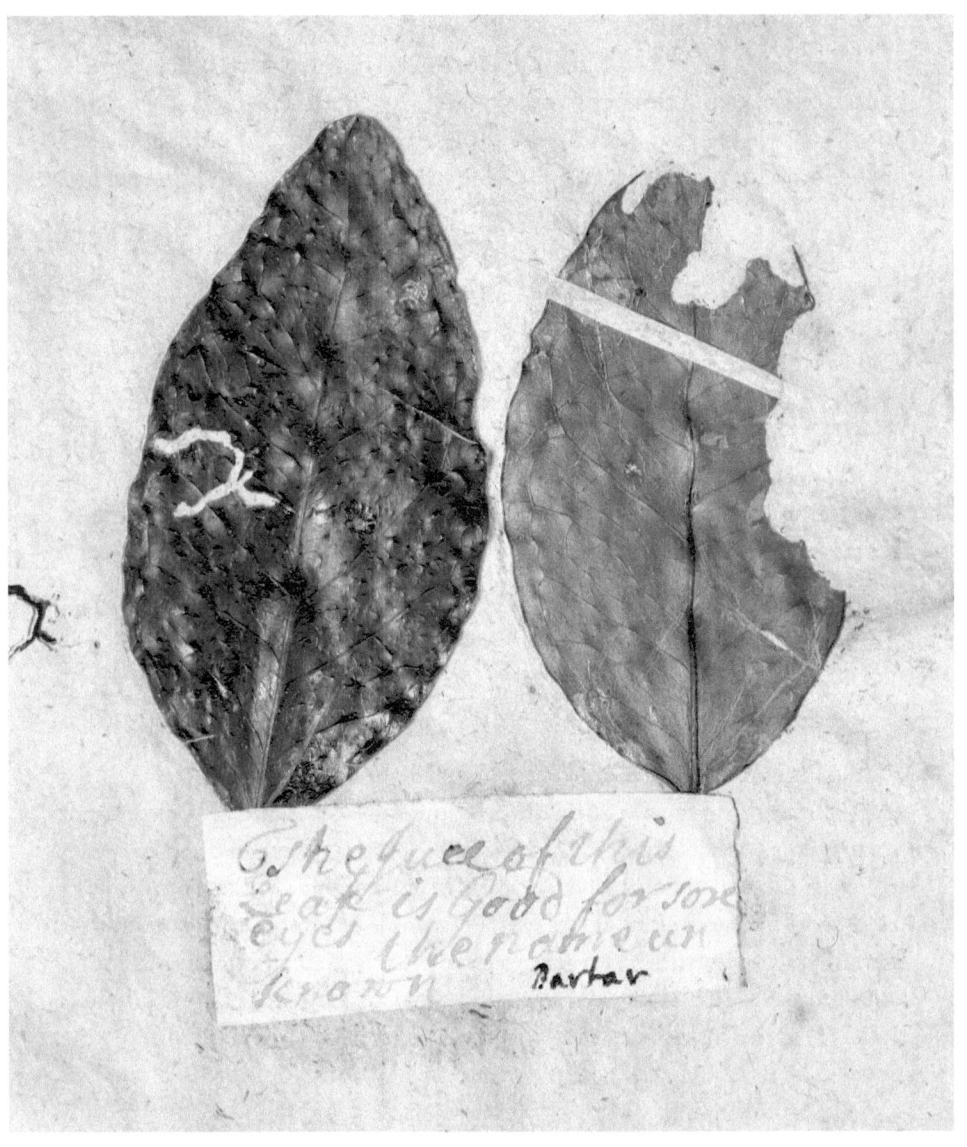

FIGURE 1.2. Herbaria specimens obtained by Edward Bartar. Two leaves sent by the Royal African Company slave trader to James Petiver in the 1690s. The accompanying label, likely written by Bartar, noted that along the Gold Coast, the juice of the leaves was known to be good for curing sore eyes. Sloane Herbarium, HS 155, f. 161. © The Trustees of the Natural History Museum, London.

likely by the Fante people.[28] Yet the label included no acknowledgment of its reliance on West African natural knowledge. In this it was similar to most collections gathered along the routes of the slave trade. The two leaves "good for sore eyes" reflected both the physical extraction of West African natural objects and the attempted erasure of the African natural knowledge that had motivated the specimens' collection in the first place.

Collections such as those gathered by Bartar at Cape Coast serve as reminders that the centrality of the Royal African Company within the late seventeenth-century English slave trade made it similarly important within networks of natural history. Metropolitan naturalists profited from England's trade with Africa through the collection of rare and often previously unknown flora and fauna. In the late seventeenth century they largely did so with the assistance of Royal African Company employees, who in turn relied upon African knowledge and collectors. While metropolitan naturalists' reliance on the routes of the slave trade continued throughout the eighteenth century, in time the Royal African Company became less central to naturalists' efforts to exploit the slave trade in order to collect Africa's natural productions.

COLLECTING FROM A SLAVE SHIP

In the winter of 1707 slave ship surgeon James Fraser visited Petiver's apothecary shop on Aldersgate Street in London. Fraser was already a veteran of the transatlantic slave trade. He had sailed as ship surgeon on the *Mayflower*, which had transported captive Africans from the Bight of Benin to Jamaica three years earlier. In January of 1707 Fraser was busy preparing for another slaving voyage. This time he was employed on a slave ship bound for Whydah and then Jamaica. His visit to Petiver's shop was likely part of these preparations. Slave ship surgeons were responsible for ensuring that their medicine chests contained everything they would need to care for crew and captives during the long voyage. Fraser's visit to Petiver's shop allowed him to purchase the various drugs and supplies he required. It also gave the apothecary a second chance to recruit Fraser as a collector of natural historical specimens. Fraser had promised to gather specimens for Petiver during his previous slaving voyage. Despite Fraser's failure to keep this pledge, Petiver remained optimistic that the surgeon might obtain natural curiosities during his next voyage. If Petiver's correspondence is any indication of the conversation between the two men, the apothecary likely chastised Fraser for failing to obtain specimens while onboard the *Mayflower* but also promised to "remember him to posterity" by acknowledging in print any specimens he brought back from his next slaving voyage.[29] Petiver must have been persuasive. A few weeks later, Fraser assured Petiver that "if God continues my health while at the Guinea coast or other places where a Collection can be made endeavours shall be used to ans[w]er your expectations."[30]

By the end of January 1707, Petiver had refilled and returned Fraser's medicine chest. Separately, he sent Fraser some of his natural historical texts to encourage him to keep his pledge. These included what Petiver called his "Materia Medica Guindarsis," a catalog of forty West African plants based on the collection from Cape Coast Castle sent by the African Company minister Smyth in 1693. Petiver also gave Fraser two copies of his *Gazophylacii Naturae & Artis, Decas Prima*, which he noted contained "several things from Guinea" as well as engravings of plants that British naturalists had not yet acquired, such as the Ethiopian long pepper and ipecacuanha. The second copy of *Gazophylacii Naturae & Artis* that Petiver sent Fraser was for the surgeon to give to anyone he recruited to join Petiver's network of collectors.[31]

Petiver had good reason to be optimistic that Fraser might "endeavor... to ans[w]er" his expectations. By 1707 the apothecary's museum already contained numerous specimens gathered on his behalf by slave ship captains and surgeons. For example, the ten images in *Gazophylacii Naturae & Artis* that Petiver sent to Fraser included eight specimens gathered in West and West Central Africa by slaving surgeons: a yellow fly from Cape Mesurado, grass from Cape Three Points, unicorn beetles from Whydah and Old Calabar, an aquatic plant from Whydah, and three Angolan butterflies.[32] Through gifts of natural historical texts, Petiver attempted to guide and encourage the collecting efforts of slaving mariners like Fraser who promised to gather specimens on his behalf.

Slaving mariners such as Fraser became metropolitan naturalists' most important collectors in West and West Central Africa in the early eighteenth century. The deregulation of Britain's slave trade and its subsequent growth brought increasing numbers of Britons, especially mariners, to coastal Africa. Slave ship captains and surgeons employed by independent traders rather than the Royal African Company were the primary collectors of African flora and fauna for British naturalists in the first decades of the eighteenth century. Reflective of this trend, Fraser's employers while he was a surgeon on the *Mayflower* in 1703 were four independent traders.[33] Maritime collectors like Fraser spent less time in West and West Central Africa than would be typical for land-based collectors employed by the Royal African Company. Yet the itineraries of their slaving voyages meant that they could gather specimens in multiple locations, both in Africa and in the New World. Further, the deregulation of the British slave trade resulted in British merchants and mariners trading in more regions of West and West Central Africa. The geographies of natural historical collecting within Africa similarly expanded to encompass most regions of West Africa and occasionally West Central Africa. The provenance of the eight African specimens in *Gazophylacii Naturae & Artis*, stretching from Cape Mesurado to Angola, reflects the broader geography traced by British slave traders in the early eighteenth century. Slave ship surgeons and captains also functioned as important nodes of connection within the transatlantic networks of natural history by delivering letters, gifts, and specimens between correspondents resident on different continents.

The potential profits to be made enticed English merchants throughout the seventeenth century to illicitly participate in the slave trade in violation of the Royal African Company's monopoly. One estimate suggests that between 1674 and 1686, one in four enslaved Africans brought to the Americas was transported illicitly by an independent English trader.[34] Yet beginning in 1698, any English merchant sailing out of any English port could legally participate in the transatlantic slave trade provided that they pay the Royal African Company a 10 percent duty for the upkeep of the company's forts in West Africa. The 1707 Acts of Union, which joined Scotland and England into one kingdom, opened the same privileges to Scottish merchants. Parliament allowed the requirement that independent traders pay the Royal African Company a 10 percent duty to lapse in 1712, thus fully deregulating the British slave trade. As historian William Pettigrew argued, deregulation "transformed the contours and capacity of Britain's slave trade."[35]

The end of the Royal African Company's monopoly contributed to the dramatic growth of the British slave trade in the eighteenth century. By the time Fraser sailed in 1707, the number of enslaved Africans transported on British vessels was more than twice that as when Petiver began to collect natural curiosities in 1690. British merchants had been the leading shippers of captive Africans since the 1670s, a trend that deregulation only accelerated. Increasing numbers of British merchants and mariners sought to profit directly and indirectly from the commerce in human beings. So too did British naturalists.[36]

The transatlantic slave trade served as a conduit, bringing African and American plants, animals, and shells into European collections. For example, the extensive museum belonging to Sir Hans Sloane benefited from the collecting efforts of British mariners likely engaged in the transatlantic slave trade. Sloane's museum included at least twenty items gathered by a Mr. Campbell, "an ingenious surgeon" whose travels brought him to Jamaica,

Antigua, and West Africa. Although Sloane does not indicate the trade in which Campbell was engaged, his itinerary suggests he was employed in the transatlantic slave trade. Campbell obtained man-made objects from Africa such as "the skin of a catt worn as an apron in Guinea," soap made of palm oil and wood ashes, a garter composed of cotton, and a piece of cloth produced from a palm leaf. Campbell's collections also included "seeds, fruits, and vegetable substances from Guinea and Antego," a remedy for epilepsy derived from wild cassava from Jamaica, Antiguan shells, Caribbean sea urchins, and fossils from both Jamaica and Guinea. Capt. William Walker contributed a mix of man-made and natural objects to Sloane's museum. His collections included "a Capashears knife" and a black scorpion from Guinea as well as a shell from the Bahamas and porcupine quills from the West Indies.[37] Sloane's museum also included what the physician described as an "Indian drum" from Virginia, which modern scholarship has identified as an Akan drum constructed out of West African woods. Slave ship captains regularly forced captive Africans to dance on deck as a form of exercise. The Akan drum in Sloane's collection was likely collected by a slave ship captain for this purpose. Today it is the oldest African American object in the British Museum.[38] Another member of the Royal Society, Dr. James Douglas, relied on specimens gathered by a slave ship surgeon to conduct his natural historical investigations. The essay on the coconut that Douglas read before the Royal Society, for example, was based on specimens gathered by British slave ship surgeon Robert Lightfoot in Barbados and São Tomé.[39]

Petiver relied extensively on British mariners to acquire a natural historical collection of global scope while rarely leaving London. More than two-thirds of the mariners who obtained specimens for Petiver in the Atlantic World were engaged in the transatlantic slave trade.[40] These twenty-five slave ship surgeons and captains collected flora and fauna, transported specimens, and recruited new collectors for the naturalist. The slave ship surgeon James Skeen typically collected specimens for Petiver in the company of another slaving surgeon, William Watts. Together, Watts and Skeen filled three or four books with pressed plants from Cape Three Points on the Gold Coast and Whydah in the Bite of Benin. These included *Adhatoda* from Cape Three Points and a Widense arum plant. The pair also acquired shells and insects from the same two slaving ports. Skeen gave Petiver a unicorn beetle, a white butterfly from Whydah, a "brown Guinea butterfly" from Cape Three Points, brown and medium-gray butterflies from Anomabu, and shells from Jamaica.[41]

British slave ship surgeon Richard Planer collected *naturalia* on both sides of the Atlantic during multiple slaving voyages stretching over five years. He gathered a yellow butterfly, finely toothed moss, ferns, and assorted other plants from Cape Mesurado on the Windward Coast. He also brought Petiver a "harmless and very beautiful" Pompom lizard, several insects from "the *Guinea* Coast" and a yellow African fly. Planer's collections included a black butterfly and "very curious Insects" from the coast of Cartagena. Petiver noted that many of the specimens he received from Planer were "not yet taken notice of" by British natural historians.[42]

The specimens Planer acquired were often stowed in the slaving surgeon's chest alongside other objects exchanged between Petiver and his terrestrial correspondents. When Planer's ship called at Cape Coast Castle, for example, the surgeon delivered letters and gifts from Petiver to the Royal African Company agent Edward Bartar. Petiver hoped that the slave ship surgeon would receive from Bartar specimens in return, which Planer could transport back to London. As Petiver informed Bartar, "I hope also you will not let Mr Planer come away empty handed" but instead would entrust Planer with a collection of "whatever plants shells & Insects you have by you & whatever more you can get before he goes away." When Planer made a second delivery to Bartar at Cape Coast Castle a year later, Petiver repeated his request that Bartar take advantage of the opportunity to send specimens by way of Planer's slaving vessel, writing, "I desire you will not fail to send by him what you have already got." In order to amass an even larger collection, Petiver recommended that Bartar send some of his enslaved people "very often abroad" to collect specimens while Planer remained at Cape Coast. He asked, in particular, for all the small lizards and snails Barter could procure. These, Petiver instructed, should be packed in small bottles or jars filled with rum, the cost of which the naturalist pledged to repay "double whatever charge you are att."[43]

The objects gathered by mariners such as Planer, Skeen, and Watts illustrate how the geographies of the slave trade powerfully molded the collections of British naturalists. The provenance of the specimens they acquired reflected the routes of the British transatlantic slave trade in the early eighteenth century. Slaving ports commonly visited by British slave ships, such as Anomabu, Cabinda, Cape Coast, Cape Three Points, Old Calabar, and Whydah, occur again and again in Petiver's and Sloane's museum catalogs and publications. Occasionally, slaving mariners also gathered specimens in American ports of disembarkation. West African specimens such as the shells, insects, and plants Skeen and Watts collected in the Bight of Benin and the Gold Coast were likely gathered during the months a slave ship typically spent along the African coast purchasing African men, women, and children.

The weeks or months a slaving vessel spent in American ports of disembarkation while the crew sold their captive Africans and acquired plantation products for the return voyage would have provided slaving mariners an opportunity to collect American specimens such as Skeen's Jamaican shells. The objects in these collections reflected the tenuousness of British presence along the African coast. Just as the British projected little if any political power along the African coast, so too most Britons collecting specimens in West and West Central Africa rarely ventured far beyond the littoral.[44] The natural curiosities that slave ship surgeons and captains presented to Petiver and Sloane were usually specimens that could be collected on board ship or within its sight. In particular, shells, coastal plants, butterflies, and other insects predominated among these collections.

As the example of Richard Planer illustrates, slaving mariners served as go-betweens within the larger transatlantic networks of natural history. Ship surgeons and captains engaged in many different types of commerce. They transported collections of natural curiosities, superintended delicate plants and seeds during voyages, and delivered letters and packages on behalf of metropolitan naturalists.[45] Mariners on British slaving vessels were a key component of this latticework of connections within the Atlantic World. Slave ship captains and surgeons loaded letters, gifts, and African and American curiosities among their personal belongings, saving metropolitan naturalists freight charges and supervising the care of specimens during their journeys to Britain. Mr. Danvers, the surgeon on the Royal African Company's vessel the *Prince George*, presented a hamper containing thirty-six bottles of "stout English beer" to Bartar at Cape Coast in 1697. The accompanying letter from Petiver urged Bartar "not [to] fail of sending by earliest shipping a collection of what Shells Insects & Grasse may come into your way." The following year Danvers delivered a collection of Gold Coast plants sent by Bartar to Petiver in London.[46] Over the course of seven years, Petiver sent Bartar gifts of natural historical texts, writing paper, collecting supplies, purging powder, beer, and newspapers—all delivered by slave ship surgeons. Other slaving mariners served as go-betweens with Petiver's American correspondents. The slave ship surgeon William Brown delivered "severall Tables of American Ferns" to Petiver's correspondent in Jamaica, Dr. David Crawford, in addition to obtaining plants while in Angola. In return, Crawford entrusted a Jamaican giant galliwasp to Brown's care. Crawford sent Petiver letters and specimens by means of at least four different slave ship surgeons over a seven-year period.[47] British slaving mariners provided some of the few direct, personal links between correspondents separated by an ocean.

Gifts offer one explanation for why British slave ship surgeons and captains collected on naturalists' behalf. Richard Planer, for example, received a copy of Petiver's *Musei Petiverani* only months after it was published. Similarly, the naturalist sent slaving surgeon James Fraser copies of his *Gazophylacii Naturae & Artis, Decas Prima* to encourage him to collect in 1707.[48] A few slaving mariners may have shared an interest in natural history and a desire to participate in the circuits of early modern science. Fraser, for example, asked Petiver's advice about whether his observations on a Bermudian herb were worthy of submission to the Royal Society. When the *Mayflower* was delayed departing Plymouth in 1703, Fraser passed the time on board by dissecting fish. He shared with Petiver his discovery that the stomachs of local cod often contained shells. For someone of Fraser's interests, gifts of natural historical texts and the opportunity to participate in the transatlantic networks of natural history might have been recompense enough. For such an individual, Petiver's frequent promise that collectors would be remembered "to posterity" in his publications might have served as further motivation.[49]

Yet Fraser was surely unusual among the slaving mariners who collected on behalf of metropolitan naturalists. The vast majority of the thousands of mariners employed in the British slave trade never collected natural historical specimens, and of those who did, it is difficult to imagine that many of the violent and desperate men described in the historiography of the slave trade were motivated by an abstract interest in natural knowledge. Therefore, it is more likely that Petiver's collectors received tangible payment to reward their efforts. Although there is no record of what compensation Petiver offered his maritime collectors, the naturalist's correspondence with terrestrial collaborators offers some hints of what these arrangements might have been.[50] Some colonial collectors received payment for specimens in kind, in the form of European newspapers, books, food, or drink. Others received medicines, presumably drawn from the stocks of Petiver's apothecary shop. In 1694, for example, Petiver sent Bartar purging powder, good for treating "Clapps and all cases where purging is convenient." In some cases, such arrangements were primarily commercial in nature. Hannah Williams of South Carolina sold the imported medicines she received from Petiver in exchange for her natural historical collections. Petiver was also willing to purchase specimens outright. Disappointed with the quantity of specimens he received from Dr. David Crawford of Jamaica, Petiver argued that Crawford could easily recruit "any common Soldier or Negro Slave once or twice a Weeke to make a collection." He offered the doctor "a Crown for every Quire you shall get fil'd for me." Similarly, the naturalist offered a correspondent in Antigua five shillings for every 4 dried plants, 10 small animals, or 100 butterflies or moths he received.[51] Even in the absence of a pre-negotiated payment scale, sailors returning from distant ports could trust that European collectors would be willing to purchase exotic items acquired during their travels. Sloane, for example, added a West African plant to his collection after a friend purchased it on his behalf from a ship captain's widow. Since the fifteenth century, objects from West Africa's souvenir market such as amulets and carved ivory horns found their way into European cabinets of curiosity. Mariners had found a ready market in London for both natural and man-made objects; slaving mariners who collected specimens on behalf of British naturalists were part of this long-standing practice.[52]

The British slaving mariners who acquired African specimens for metropolitan naturalists in the early eighteenth century appear to have all been ship surgeons or captains. Captains' and surgeons' social status, shipboard responsibilities, and training explain why British naturalists typically entrusted them with their biocargo and relied on their efforts as collectors. Moreover, these men tended to spend more time on shore as they were usually responsible for determining which of the captive Africans offered for sale would be added to the ship's human cargo. The time slave ship surgeons and captains spent on shore meant that they had a greater opportunity to acquire specimens than most of the ship's crew.[53]

A captain's absolute authority over the people and physical spaces of his vessel made his cooperation critical. Given the delicate nature of many natural history specimens, their precise placement on board the ship could determine whether they survived the voyage intact. Improper levels of sunlight, moisture, and heat, or exposure to a vessel's many rats, cats, and other pests, could easily destroy the pressed plants, pinned butterflies, and other specimens on which natural history depended.[54] Decisions regarding where collections would be stowed or even if they would be allowed on board ultimately lay with the vessel's captain. Petiver, at least, also encouraged the captains among his acquaintance to employ their shipboard authority in order to gather additional *naturalia*. In 1716 the naturalist requested that Capt. George Jesson "lend my flycatchers to some of your blacks whilst your on the Island [of Jamaica]" in order to "take & kill whatever butterflies & Moths they meet." Although Petiver's precise meaning is unclear, it raises the possibility that he hoped Jesson would employ some of his captive Africans as collectors.[55]

More than twice as many ship surgeons than captains obtained specimens for Petiver along the routes of the British slave trade.[56] The frequency with which transatlantic slave ships carried a ship surgeon made them unique among the commercial vessels plying the Atlantic basin. British slave ships were not required to carry a ship surgeon until the passage of Dolben's Act in 1788. Yet doing so was common practice long before it was law. Contemporary accounts and modern scholarship agree that death and disease plagued the slave ship. The practice of including a surgeon among a slave ship's crew developed in response to this grim reality. By employing surgeons, ship owners and captains hoped to lessen the gruesome mortality among captives and crew and thereby increase their profits.[57] Petiver's apothecary practice likely created opportunities for him to recruit slave ship surgeons as potential collectors. Slave ship surgeons like Fraser may have visited Petiver's shop to purchase the medical supplies they would need during the transatlantic slaving voyage. As historian Katrina Maydom has demonstrated, Petiver had a habit of recruiting customers to become collectors.[58]

Slave ship surgeons occupied the lowest rung within London's medical hierarchy, but they would have claimed one of the highest levels of education among a typical British slaving crew.[59] In theory, at least, this education would have emphasized *res naturae*, the things of nature. Surgeons who trained in Edinburgh, for example, would have learned botany from James Sutherland, professor of botany and intendant of the physic garden at the University of Edinburgh. Edinburgh's Incorporation of Surgeon-Apothecaries paid Sutherland a guinea per student to lead the group's apprentices on quarterly botanizing expeditions and to instruct them in botany in the physic garden.[60] British slave ship surgeons were expected to have a basic understanding of anatomy, surgery, and physic. Like physicians whose extensive contributions to natural history have been more thoroughly studied, ship surgeons had a professional interest in the exotic minerals, animals, and plants encountered in their journeys, particularly those with medicinal properties.[61]

The slave ship surgeon required a discerning eye. Along with ship captains, slaving surgeons needed to be able to distinguish healthy bodies from those exhibiting signs of sickness. The slaving surgeon and captain examined captives offered for purchase along the African coast, determining whether "they are physically fit, have healthy eyes, good teeth, stand over four feet high, and if men, are not ruptured; if females, have not 'fallen breasts.'" Ship owners expected that surgeons in their employ could, after a quick inspection, determine whether captive Africans were "likely" or whether their bodies concealed hidden sicknesses or infirmities. They expected surgeons and captains to see health and sickness at a glance, and to discern relative value and potential profits rather than humanity in the captives before them. In essence, ship owners expected them to have a collecting gaze.[62] The collecting gaze required of the slave ship surgeon was not very different from the collecting gaze of a virtuoso such as Petiver or Sloane. As historian of science Lorraine Daston noted of descriptive sciences such as botany and anatomy, "these were the sciences of the trained eye, accustomed by years of experience to distinguish the essential from the accidental, the normal from the pathological, the typical from the anomalous." Like the naturalist, the slave ship surgeon's "task was to extract the truths of nature from the welter of confusing appearances."[63]

Petiver's and Sloane's correspondence, publications, and museum catalogs suggest that some British slaving surgeons and captains employed a collecting gaze during the months they spent along the African coast, gathering both natural curiosities and African captives. In the early eighteenth century, metropolitan naturalists exploited Britain's growing engagement with the transatlantic slave trade to obtain specimens from West and West Central Africa with the assistance of slave ship surgeons and captains employed by independent merchants. Slaving mariners' time on the African coast was brief, but the itineraries of their voyages enabled them to acquire specimens at multiple slaving ports on both sides of the Atlantic. British slaving mariners collected in West and West Central Africa at a time when few other Britons were in a position to do so.

"HAVING AS YET VERY FEW ... FROM THAT PART OF AFRICA"

In 1695 James Petiver was unsure of the whereabouts of his correspondent John Smyth. He thought that the Royal African Company minister should have long departed Cape Coast, and he had been in "great hopes" of seeing him in London. Surmising that Smyth might still be in West Africa, the apothecary sent one last letter to the Anglican minister by means of a slave ship surgeon bound for the Gold Coast. Petiver thanked Smyth for a collection of African plants that he had sent in 1693 but also confided that a second collection Smyth intended for Petiver had never made it to England. Although Petiver was uncertain whether the chaplain was still at Cape Coast Castle, he was not one to turn down a chance—however slight—to obtain more specimens. Petiver therefore reminded Smyth that he would gladly welcome any further collections he made before departing West Africa. "Rev'd Sr," the apothecary pleaded, "We having as yet very few either shells or none Insects from that part of Africa where you are I must humbly beg you will be pleased to have some Native or poor man to make me what Collections of these and Plants may come in his way." If such a collection was sent "by the very first Shipping," Petiver predicted he would "want words to express how acceptable" it would be.[64]

The work of early modern naturalists required that they consult, compare, and analyze natural historical specimens or their visual representations. As Petiver informed Smyth, there were few such specimens from West Africa for British naturalists at the turn of the eighteenth century to consult. Naturalists therefore enthusiastically welcomed—and relied upon—the flora, fauna, minerals, and other objects obtained by British slave ship surgeons, captains, and slaving agents in the late seventeenth and early eighteenth centuries. The group of British slaving mariners and agents engaged in collecting natural historical specimens was likely small, yet their impact on British natural history was significant. Slave traders and mariners collecting on behalf of British naturalists, in turn, relied on Africans as guides, informants, collectors, and knowers. The uncertain fate of Smyth's second collection from Cape Coast Castle reminds us that metropolitan naturalists' reliance upon the routes of the slave trade as a site of natural history did not always go according to plan. Yet when it did, these collections profoundly shaped the production of natural knowledge. Natural objects and knowledges extracted by means of the British slave trade were springboards for discussion among naturalists, appeared in scientific texts, and joined natural historical museums.

A specimen gathered by a Fante in the Gold Coast, then acquired by a British slaving mariner, transported on a slaving vessel to an American port of disembarkation and then to London, might find its way into the collection of a metropolitan naturalist and eventually into the Royal Society's meeting rooms in Gresham College. There, naturalists such as Petiver and Sloane presented the specimens and observations obtained by means of the British slave trade to the assembled members of the learned society. In December 1698, for example, Petiver "produced a Strange leaf from Calabar in Guinea" that was "perforated after a very odd manner." The following year he displayed the bill of the African spoonbill and a cashew nut "from Guinea." Later that same year Sloane reported that "Cap't Forty" (likely Capt. Henry Forty of the slave ship *Neptune*) described the mouth of the River Gambia as "accessible and wide Enough, without Shoal, for a Ship to go up, and he added that in the Country thereto belonging, there are Ostriches." Petiver regularly presented specimens he obtained from slaving mariners during meetings of the Royal Society, often elaborating on the descriptions he provided in his published accounts of the same objects. For example, he noted that the specimen of "Bartars dark Guinea Butterfly with white spotts" displayed before the society was "of the same magnitude and delineation" as the one he had published in *Gazophylacii Naturae & Artis*. The naturalist added that the wings were transparent "if held to the light" and that the body "sparkles with white on a black ground."[65]

Specimens gathered along the routes of the slave trade were regularly shared, bartered, and borrowed among British naturalists. The English physician and naturalist Martin Lister, for example, borrowed African specimens from Petiver's collections to complete his masterpiece on conchology, *Historiae Conchyliorum*. The botanist Leonard Plukenet described Gold Coast plants gathered by Bartar in his *Almagestum Botanicum*. Similarly, William Sherard's herbarium included a plant gathered by Bartar near Cape Coast Castle even though there is no evidence that Sherard directly corresponded with the slaving agent. The first Sherardian Professor of Botany at Oxford, Johann Jacob Dillenius, consulted Gold Coast plant specimens gathered by Bartar as well as specimens collected by slaving surgeons Skeen, Kirckwood, Planer, and Mason in the course of his botanical research. Specimens gathered along the routes of the British slave trade continued to circulate among and be consulted by British naturalists long after the means of their collection were forgotten.[66] Herbaria specimens sent by Smyth, for example, were used in a recent study comparing historical and contemporary ethnobotanical knowledge of Ghanaian plants.[67]

Specimens obtained by means of the British slave trade were also described and depicted in metropolitan naturalists' publications. Early modern naturalists regularly relied on images of specimens in the course of their studies. The illustrations in Petiver's texts were modest compared to the lush engravings found in other natural histories of the period.[68] Yet they still allowed European naturalists to study specimens that they likely would never view in person. Decades after Petiver's death, Swedish systematist Carolus Linnaeus referenced illustrations from the apothecary's publications in order to classify various African species. For example, slaving surgeon Richard Planer obtained a West African moth "from the Guinea Coast" around 1700, which he gave to Petiver. Petiver described the moth as "*Phalaena* Guineensis flava perelegans & pulchre oculata" and included an image of it in his *Gazophylacii Naturae & Artis: Decas Tertia*, published in 1704 (fig. 1.3). Half a century later, Linnaeus used this image to define the species he called *Phalaena paphia*. Through the circulation of images, a specimen gathered by slave ship surgeon in West Africa could serve as the basis for taxonomical study decades later. In similar fashion, objects gathered by British slaving mariners and agents shaped the development of natural inquiry throughout the eighteenth century and beyond.[69]

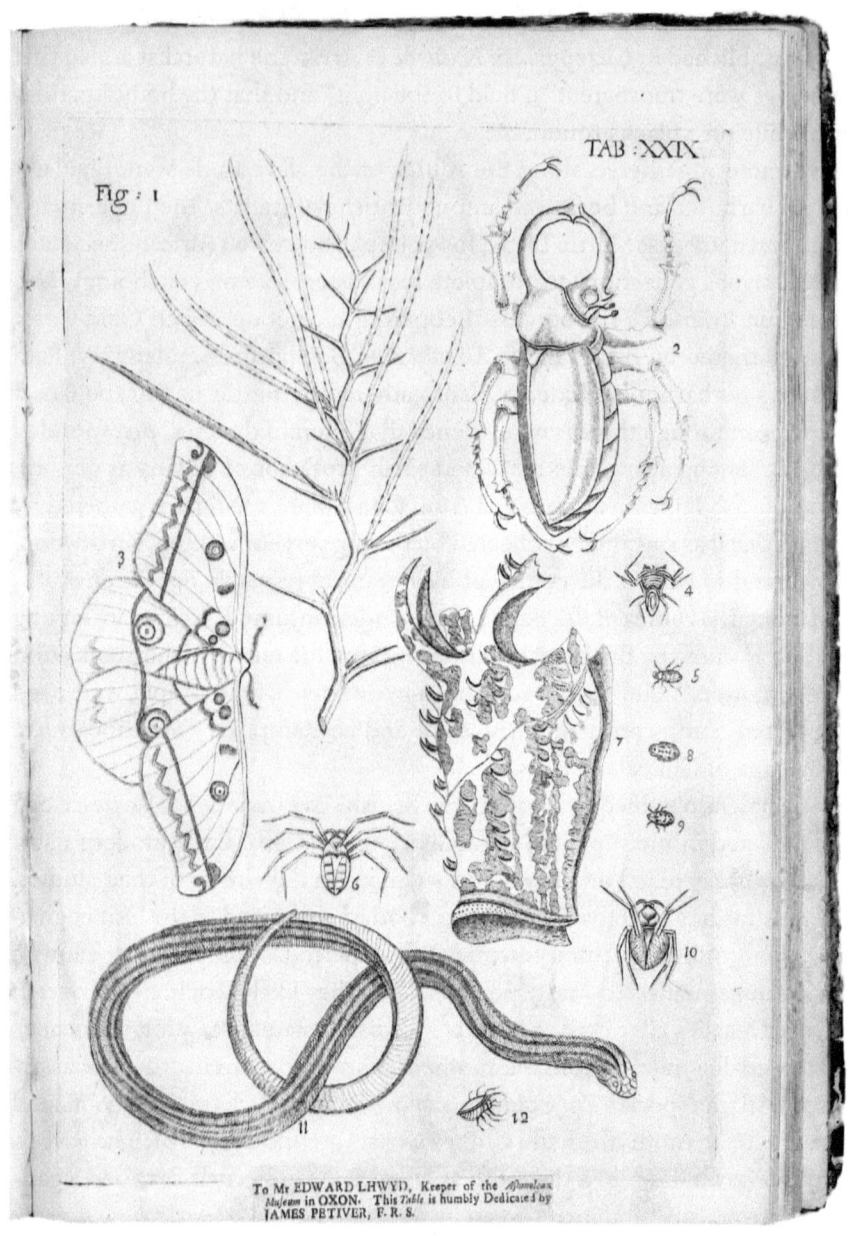

FIGURE 1.3. *Phalaena guineensis*. James Petiver featured the West African moth labeled figure 3 on tablet 29 of his *Gazophylacii Naturae & Artis: Decas Tertia* (1704). Petiver received the moth from slave ship surgeon Richard Planer around 1700. Decades later Carolus Linnaeus described the *Phalaena paphia* based on Petiver's published image. James Petiver, *Gazophylacium Naturae & Artis Decas Prima-[Quinta]*, London (1702–6), RB 497024a, Huntington Library, San Marino, California.

West African specimens obtained by means of the British slave trade played an important role in establishing Petiver's reputation as a naturalist. The collections that Smyth sent from Cape Coast Castle in 1693 became the basis for Petiver's first contribution to the Royal Society's journal, *Philosophical Transactions*. "A Catalogue of some *Guinea-Plants*, with their *Native Names* and *Virtues*; Sent to *James Petiver*, Apothecary, and Fellow of the Royal Society" appeared in 1697, just two years after Petiver's election to membership in the Royal Society. The society promoted itself in the late seventeenth century as a clearinghouse for useful knowledge and aspired to catalog the world's natural productions. "Catalogue of some *Guinea-Plants*" thus provided a means for Petiver to prove himself worthy of his election by publishing descriptions of over three dozen Gold Coast plants at a time when specimens from West Africa were difficult for English naturalists to obtain.[70]

As the title indicated, "Catalogue of some *Guinea-Plants*" indexed West African plants and "their *Native Names* and *Virtues*." Enslaved or free Fante peoples living along the Gold Coast likely selected, gathered, and preserved the forty specimens described in the text. Each entry began with the name by which Smyth reported the plant was known along the Gold Coast. It then briefly described the preparation and use of medicaments derived from the plant. Most of the entries included additional information such as Latinate classifications, citations to natural historical or medical texts, or morphological descriptions.[71] Each entry ended with Smyth's name or initials. For example, the first entry explained that *aclowa*, "so called by the Natives in *Guinea*, dried and rub'd on all the Body is good for the Crocoes (or Itch.) Mr. *John Smyth*." Through the repetition of Smyth's name and initials, Petiver sought to remind readers that his informant was the Royal African Company minister. He thus attempted to position Smyth as the author of the knowledge compiled in "Catalogue of some *Guinea-Plants*," rather than the Fante people living along the Gold Coast, whose natural and medical knowledge likely most shaped the text. The very organization of the catalog, however, emphasized West African knowledge. Whereas most natural history texts of the period arranged specimens by their Latin polynomials, "Catalogue of some *Guinea-Plants*" was arranged alphabetically by the local names by which the plants were known along the Gold Coast.[72] The text's structure and emphasis on Fante names and knowledges partially ceded intellectual authority back to the African knowers who likely were the sources of the medical and botanical knowledge in the first place. Such an organization also potentially made the text more useful for Britons in the Gold Coast as a guide to local medicaments.[73]

"Catalogue of some *Guinea-Plants*" operated on two registers, which corresponded with the two titles by which the text was known. The text's published title reflected its function as a catalog to a collection of botanical specimens that in 1697 was in Petiver's physical possession. The collection included forty-six pressed plants and handwritten labels sent by Smyth from Cape Coast Castle. Forty of these plants were described in the published catalog. "Catalogue of some *Guinea-Plants*" included hints about the appearance of particular specimens in Petiver's museum. Petiver noted, for example, that the leaves of the *aconcroba* were stiff, opaque, and black because they were "now dry." The text described how Gold Coast plants appeared after having been pressed, dried, and transported to his home on Aldersgate Street in London. All but two of the Cape Coast specimens sent by Smyth in the 1690s survive today as part of the Sloane Herbarium in the Natural History Museum in London. The herbaria sheets materially bind the published catalog with the plant specimens. Petiver used the published "Catalogue of some *Guinea-Plants*" as the source of his specimen labels, as shown in figure 1.4. The naturalist pasted the published description of each plant cut from the catalog onto the physical herbaria specimens, thus joining text and specimen onto the same page. A few surviving specimens also include labels in Smyth's handwriting.[74]

The title Petiver himself most frequently used for the text—his African or Guinea *materia medica*—framed it as a practical guide to medical substances along the Gold Coast.[75] Petiver had a professional interest in West African drugs as an apothecary and would have known well the growing global market for medical substances.[76] Further, a "Guinea *materia medica*" served the interests of British commerce and colonialism at a moment when British slaving along the West African coast was expanding rapidly. "Catalogue of some *Guinea-Plants*" promised relief to the perennial problem facing Britons in West Africa: how to stay alive in the "white man's grave."[77] Although many places in the early modern period had a reputation for having an unhealthy climate for Europeans, West Africa was among the most deserving of such a reputation.[78] "Catalogue of some *Guinea-Plants*" reflected the principle that one should use local plants to cure local diseases, as well as the pragmatic interest in local healing techniques and knowledges that frequently characterized the behavior of individuals navigating a new or deadly disease environment.[79] To do so, the catalog relied upon Gold Coast natural knowledges. The text noted, for example, that in Cape Coast, local people used *acroe* for recovering one's strength, *aconcroba* to cure smallpox, and *concon* for killing worms in the legs. "Catalogue of some *Guinea-Plants*" informed

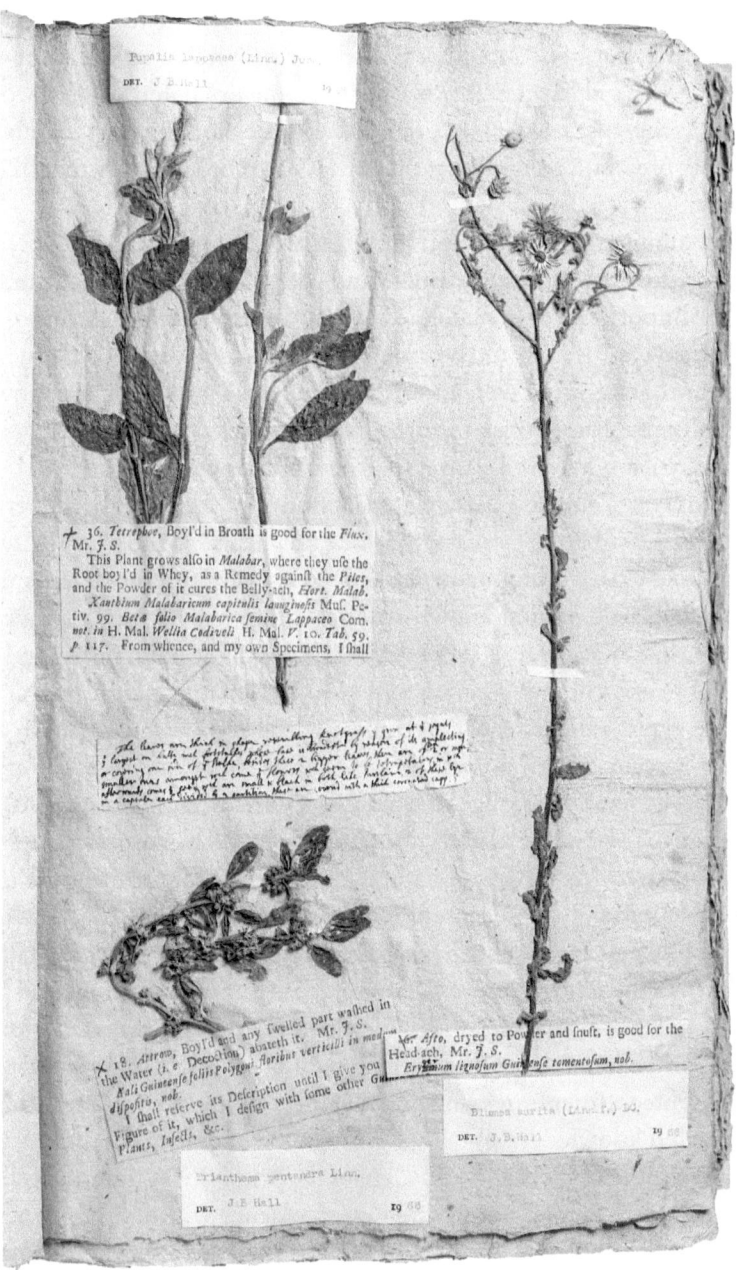

FIGURE 1.4. Herbaria specimens obtained by John Smyth. Volume 191 of the Sloane Herbarium contains the surviving forty-four Cape Coast plant specimens sent by John Smyth to James Petiver in the 1690s. The specimens bear Petiver's late seventeenth-century labels, which were cut and pasted from his published description of the plants, and twentieth-century labels were added by J. B. Hall in 1966, which provided the modern botanical identification for each plant. Sloane Herbarium, HS 191, f. 2, © The Trustees of the Natural History Museum, London.

readers not just which plants to use but also how to use them. For example, it reported that *aconcroba* should be boiled in wine but that *concon* should be pounded and mixed in oil.[80] It explained that *afto* dried and snuffed would cure a headache, that *assrumina* "pounded and rub'd on the Legs, kileth the Worms that breed there," and that *atanta* added to broth gave strength to a sick person.[81]

Petiver simultaneously desired and dismissed the African knowledges that he urged British slaving mariners and agents to collect. The naturalist framed the importance of "Catalogue of some *Guinea-Plants*" in terms of the "many Advantages" that would be brought "to the Art or Mystery of Physick, if the *Vertues* of all *Simples* were more nicely inquired into, or better known." For naturalists such as Petiver, the process of "better knowing" the virtues of West African plants involved abstracting them from their social, spiritual, and political contexts. Further, Petiver understood that the healing properties of the plants described in the catalog were known among the peoples living along the Gold Coast. He framed West African knowledge, however, as something less than true natural knowledge, declaring that Africans' "innocent Practice consists of no more Art than Composition." The naturalist suggested that the Fante had not improved upon the God-given therapeutic properties inherent to plants and, instead, that Fante medical knowledge was limited to rudimentary composition. In actuality, the ethnobotanical systems that developed over generations along the Gold Coast were highly sophisticated and well adapted to their material and cultural contexts. Petiver eagerly sought to coopt and abstract West African natural knowledge while at the same time asserting that it merely represented "innocent" or unknowing practice.[82]

The Gold Coast plant specimens and descriptions that Petiver obtained from Smyth in the 1690s became the basis for Petiver's publications and presentations to the Royal Society.[83] They helped to establish his reputation as a collector and a naturalist. Other British naturalists in the late seventeenth and early eighteenth centuries similarly relied upon specimens, observations, and abstracted knowledges obtained through the routes of the slave trade. The value of these objects and knowledges lay primarily in their rarity. Just a few years before Petiver began his correspondence with John Smyth, the British natural philosopher Robert Boyle bemoaned the difficulty of obtaining specimens from West Africa. Exploiting the commercial networks and infrastructure of the British transatlantic slave trade allowed British naturalists such as Petiver to obtain some natural historical specimens, although acquiring a particular one—like the "specifick medicines" in which Boyle was so interested—remained a challenge.[84]

CONCLUSION

The slave trade brought hundreds of specimens from West and West Central Africa into British museums, gardens, herbaria, and scientific texts in the early eighteenth century, even though such collecting activities likely only occurred on a minority of slaving vessels. British slaving agents and mariners acquired seeds, pressed plants, shells, butterflies, and other natural objects at British slaving factories in West Africa, from transatlantic slaving vessels trading along the African coast, and in British American ports of disembarkation. To do so, slave traders often relied upon the knowledge and labor of African guides, healers, hunters, collectors, and knowers. Natural historical specimens obtained along the routes of the slave trade bear traces of the commerce that made possible their acquisition, particularly the trade's distinctive geographies and personnel. As the British transatlantic slave trade expanded and opened to all British traders in the early eighteenth century, collecting continued to be integrated into the business of enslaving. Eighteenth-century natural history was built in part on the exploitations of the transatlantic slave trade.

Most slaving mariners who collected natural historical specimens were likely paid for the objects they obtained. These payments reflect the commodification of natural historical collecting. For the directors of the Royal African Company, the commodification of African natural and medical knowledges represented one strategy for finding new ways to profit through their trade with West Africa. African natural productions, especially mineral and medicinal resources, were valuable commodities in their own right. As we'll see in greater detail in the next chapter, the directors of the Royal African Company had periodically turned to natural history throughout the company's history in hopes that it might help them identify ways to revive its fortunes. Efforts to employ natural history took on new urgency after the African Company lost its monopoly at the turn of the eighteenth century. These efforts reached their zenith in the 1720s under the leadership of James Brydges, the Duke of Chandos.

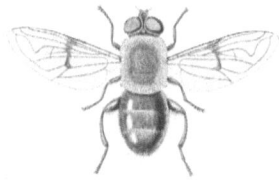

2.

Collecting for the Company

Capt. William Gower commanded the Royal African Company's vessel the *Clarendon* as it made its way to Cabinda and then to Jamaica to sell its cargo of captive Africans in 1723. While along the African coast, the *Clarendon* stopped at the company's local headquarters at Cape Coast Castle. There the castle's commander, Capt. Gen. John Tinker, entrusted Gower with a package addressed to the company's most powerful investor, James Brydges, 1st Duke of Chandos. The package contained at least twenty-one specimens of plants. The specimens were gathered near Cape Coast Castle or in the vicinity of the company's other factories along the Gold Coast. The collection included *nunum*, used by pregnant women to nourish the womb; *cussruala*, used to cure sores or boils; and *empaidoom*, used to cure pain in teeth or gums. The accompanying botanical descriptions indicated whether a given specimen was found along the coast or further inland, as well as how locals used and prepared it. *Cussuala*, for example, was pounded upon a stone until the consistency of a paste and then applied outwardly, while *empaidoom* was prepared by boiling the plant's root and frequently rinsing the mouth with the resulting liquid while it was still warm.[1]

Like the similar collection sent by the Reverend John Smyth from Cape Coast Castle thirty years earlier, the plants in the *Clarendon* collection were selected for inclusion based on African natural knowledge. West Africans, most likely Fante, had previously studied the twenty-one plants and determined how they could be used to heal, perfume, and nourish those living along the Gold Coast. It is unlikely that Tinker personally prepared the collection transported on the slave ship. Whoever did so—whether African or European, enslaved or free—used West African natural knowledge to guide their collecting efforts.[2] The *Clarendon* collection differed from that sent three decades earlier because it had been gathered on the orders of a slave-trading company rather than at the request of a metropolitan naturalist. Chandos and his allies among the company's leadership believed that collections such as the plants transported on the *Clarendon* would help the Royal African Company identify new sources of profit. They hoped that natural history could enrich the company's bottom line.[3]

By the 1720s the Royal African Company was in need of a new path to profitability. The company's 1672 royal charter had given it a monopoly on all English trade with West Africa. The monopoly was difficult to enforce from the beginning, however. Approximately one-quarter of enslaved Africans brought to English America between 1674 and 1686 were transported by independent slave traders, in clear violation of the company's monopoly.[4] The Royal African Company's position became even more precarious in the aftermath of the Glorious Revolution. No longer could it rely on royal prerogative to justify its monopoly. The company needed Parliament to provide statutory support for its exclusive trading privileges. Independent slave traders, however, lobbied for the African trade to be opened to all merchants. The 1698 statute ultimately passed by Parliament was a compromise. It recognized the African Company's royal charter, but it opened the African trade for thirteen years to anyone who paid a duty toward the upkeep of the company's African forts. Further, Parliament allowed the 1698 statute to lapse in 1712 and thereby fully deregulated the British trade to West Africa. The Royal African Company's market share within the British slave trade fell from 97 percent in 1687 to only 4 percent by 1720.[5]

Despite the Royal African Company's loss of its monopoly, a handful of its investors and officials believed that the company could find new ways to turn a profit. Foremost among this group in the early 1720s was Chandos. The duke and his allies on the company's governing body, the Court of Assistants, concluded that Parliament was unlikely to restore the company's monopoly. Consequently, enslaving would never be a source of profit. As Chandos explained to one of the company's factors in West Africa, "the [slave] Trade is upon such a foot at present from the interfering of the Interlopers, that it is not to be carried on . . . with any Advantage."[6] However, Chandos believed that trade in Africa's natural commodities such as drugs, dyes, and minerals could usher in a new period of profitability for the African Company. Surveying the natural resources of West Africa would allow the company to identify those natural commodities upon which a profit could be made. The company's Court of Assistants instructed its slaving agents to search for botanical and mineral resources indigenous to West Africa and to send specimens like those transported on the *Clarendon* to London for further testing. They also encouraged slaving agents to experiment with the cultivation of plantation products such as cotton and indigo. Their plans relied upon the appropriation of local African knowledge; the commercial infrastructure of the slave trade; the collecting efforts of Royal African Company ship captains, surgeons, and agents; and the assistance of metropolitan naturalists. Through such efforts, the duke declared that he did "not despair of having Africa become as beneficial to England as America is to Spain."[7]

"TO MAKE WHAT DISCOVERIES THEY CAN"

Chandos and his allies on the Court of Assistants believed that the slave trade was "a Losing Trade" but that better knowledge of West African drugs, dyewoods, minerals, and other natural commodities would reveal new commercial opportunities. Accordingly, the Royal African Company frequently urged its agents in West Africa "to make what discoveries they can" about African flora, fauna, and minerals. While the company's leaders used the language of "discovery," what they described can better be understood as extraction and appropriation of African knowledge.[8] Chandos and his allies urged the company's agents to study the drugs and dyes employed by local Africans, to search for sources of gold, to identify other natural commodities, and to obtain samples of all they found. They sought to enlist collecting and natural knowledge to reveal the "sure and Considerable advantage to be made" in West Africa.[9]

Chandos built his career on finding an advantage wherever he could. He made his immense fortune through a series of questionable financial dealings as paymaster general for British forces during the War of Spanish Succession. Perhaps more notorious than how he acquired his money were the ways in which he ostentatiously displayed it through his palatial estate, his collection of paintings and manuscripts, and his patronage of the arts. The duke also invested his substantial wealth into dozens of corporations and projects, including the South Sea Company, the Sun Life Insurance Company, and the York Buildings Company. His investments included schemes for improved methods of mining and refining ores, steam-powered engines, and machines to clear foul air from mines.[10]

As historian of science Larry Stewart observed, Chandos had a "passion for things mechanical, especially if they might turn a profit." The duke's interest in the practical, and profitable, application of natural knowledge was nurtured through his active participation in the Royal Society of London, to which he was elected fellow in 1694. Chandos's passion for the commercial and utilitarian applications of natural philosophy influenced his involvement with the Royal African Company. The duke became a member of the Court of Assistants and a principal investor in the company by 1720. Over the next few years he searched for ways that the application of natural knowledge might help the African Company become more profitable.[11]

Profitability for the Royal African Company, Chandos was convinced, required that its business model dramatically be revised. As historian Matthew David Mitchell argued, Chandos and his allies among the company's leadership believed that the core of its business should be the "out & home" trade that exchanged British manufactured goods for African natural commodities, rather than the slave trade. "By shifting the RAC's focus to the Europe-to-Africa route," Mitchell explained, "Chandos and his fellow Assistants believed they could make the company more profitable than it had ever been in its days of concentrating on the transatlantic triangle trade." To do so, the company would need to expand its inland trade as well as identify natural commodities close to its coastal forts suitable for the "out & home trade." Despite the long history of commercial interest in West African natural commodities, Chandos and his allies believed that the direct trade's potential had barely been tapped. Their plans hinged on the idea that West Africa contained many natural commodities yet to be exploited.[12]

A faith in undiscovered and underexploited natural resources was a common one among European naturalists, imperial officials, and merchants in the eighteenth century. Such an assertion numbered among the justifications for European colonialism. Natural history, especially fields such as economic botany, sought to identify, describe, and exploit the commercial potential of a region's flora and fauna. Naturalists enabled commercial and imperial expansion by appropriating local natural knowledge, surveying natural resources, identifying new uses for local plants and minerals, and facilitating the introduction and acclimatization of foreign botanicals. The collection of natural historical specimens and knowledge both facilitated extractive colonialism and itself was a form of it.[13]

Chandos and his allies on the Court of Assistants believed that West Africa was home to many profitable yet previously overlooked natural commodities. They encouraged the company's agents to cast their nets broadly and to bioprospect in the vicinity of the company's slaving factories. They instructed slaving agents to search in particular for drugs, dyes, and spices. Chandos frequently speculated that wherever one valuable plant was found, many others surely must grow nearby.[14] Chandos and his allies sought to locate natural commodities typically imported into Europe such as cardamom, tamarind, cinnamon, cochineal, cocoa nuts, cassia pistole, sesame, and pepper. The duke requested samples of palm oil from Whydah, India pepper from the Bight of Benin, ebony from Gambia, rhubarb from Cape Coast Castle, and cinnabar from Sierra Leone.[15] Chandos often reminded the company's agents in West Africa about the potential markets that natural commodities could expect to find in England. He noted, for example, that tamarinds and cassia pistole were reported to grow in the vicinity of Cape Coast Castle and were "Commodities which yield a very good price" in Britain.[16]

Chandos and his contemporaries were not the first to believe that West Africa contained perfumes, woods, spices, drugs, and dyes upon which the African Company could turn a tidy profit. Interest in expanding the company's trade in natural commodities tended to spike at moments when the company's primary commercial endeavors in enslaved Africans and gold seemed in jeopardy, such as in 1698. In the same letter that informed agents of the company's loss of its monopoly, the Royal African Company instructed them to search for natural commodities that could be a new source of profit. Company leaders instructed the agent at Sherbro that Parliament's decision meant that "now you must be civil toward private traders, as the Act of Parliament directs, yet it is our order that you give them no help or assistance." Simultaneously, they instructed him to send "good men up in the Country to gett all the Cam Wood you can" because, they explained, the dyewood would sell in great quantities. Agents at each of the company's forts were instructed to gather specimens of any roots, tree bark, leaves, or anything else used by locals as medicaments, dyes, or paints.[17]

Like company officials who had preceded him, Chandos especially desired the precious mineral that inspired the European name for the coast. His experience as an investor involved with various schemes to apply natural knowledge to profitable ends made him interested in how the application of natural knowledge might help the company establish its own

African gold mines. For example, in 1720 he provided Capt. Gen. James Phipps directions for a new method of determining the weight of gold dust submerged in water, which he argued would more accurately and more quickly determine its value. He urged Phipps "to try what experiments you can" using the improved method.[18] Chandos also consulted with English mining experts and veterans of the company's factories in Africa to better understand where gold might be found. Former company agents who spent time on the Gold Coast assured him that gold dust washed down from the neighboring hills after every heavy rain. The duke speculated that the source of the gold dust must be very close to the company's settlements "as the weight of Gold is specifically heavier than that of Water," therefore "no torrent whatever, unless on a very steep dissent, could carry the Particles ... any great length."[19] Chandos's consultations with British mining experts led him to conclude that deposits of gold in West Africa were "Lodg'd pretty much in the earth after the Manner of our tin." He consequently recruited a group of British tin miners to locate and mine the gold that he was confident could be found near Cape Coast Castle. Should this fail, he instructed the miners to teach the castle's enslaved Africans the more efficient method of washing gold employed in Cornwall and Devon using a device known as a buddle, "whereby one or two will wash as much in a day as an Hundred can by [the current] method." Chandos speculated that the company could improve its profits in the slave trade by employing captive Africans held at the company's forts to wash for gold until they were sold to transatlantic slave ship captains.[20]

Drugs came right after gold among the African commodities that Chandos believed would reverse the company's fortunes. He declared that West Africa "must have a great Plenty" of medicinal plants that could be profitably commoditized.[21] Chandos and his allies were confident that Royal African Company agents could identify both new drugs and local West African sources for drugs already familiar but imported from other regions of the world. They expected that African Company agents could identify local sources of cardamom, ginger, bezoar stones, ipecacuanha, rhubarb, dragon's blood, cinchona bark, and other medicaments.[22] For example, Chandos declared to Captain General Phipps at Cape Coast Castle in 1721 that it was "very probable" that the drug known as dragon's blood could be found nearby as he believed it originally came from Africa. The duke explained that dragon's blood derived from a gum that sweats off the dragon tree; the gum would be dried in the sun and then ground into a red powder. If Phipps could not locate the dragon tree, Chandos suggested that perhaps the gum of the palm tree would be a good substitute. Chandos confidently declared that any tropical drugs not indigenous to West Africa could be successfully introduced and acclimatized at the company's factories. He told Phipps that "certainly all the Druggs and plants which grow in the Southerly plantations will thrive and grow equally well with you."[23]

The Court of Assistants' interest in medical substances was rooted in the hope that new drugs and therapeutics might improve the profitability of the Royal African Company's slaving voyages. The high morality common on slaving voyages and in slaving factories along the African coast reduced the potential profits of slaving ventures. Commercial concerns therefore gave new urgency to epidemiological questions. Historians of medicine have shown how the treatment of disease and the development of practical knowledge about healing became a particular focus of natural inquiry along the routes of the slave trade during the early modern period.[24] In the British context, historian Larry Stewart has shown how the Royal African Company was quick to embrace ways that medicine might improve its bottom line by reducing mortality. In 1722 the company instructed its agents in Africa to inoculate captive Africans in response to high mortality rates on their slave ships and at their factories caused by outbreaks of smallpox. The assistants' instructions declared optimistically that the method of inoculation had "begun to be practic'd in England and hath not fail'd to succeed in every Instance." In actuality, high-profile deaths prompted fierce debate in England in the 1720s over the safety and morality of inoculation. Such concerns were not reflected in the company's instructions to inoculate all Africans trafficked by its agents.[25] More broadly, deadly disease environments along the routes of the slave trade spurred efforts to borrow, adapt, and integrate new healing techniques and drugs.

Drugs were also valuable commodities in their own right. Early modern pharmacopeias dramatically expanded over the course of the sixteenth, seventeenth, and eighteenth centuries as colonization and global trade increased the variety and quantity of imported drugs for sale. In Britain the volume of imported drugs expanded dramatically during the seventeenth and early eighteenth centuries. Chandos understood that British consumers increasingly sought foreign drugs and were often willing to pay high prices for them.[26] He calculated that if the company's enslaved Africans gathered the valuable plants and roots, drugs would be "very Cheap and cost Little or nothing. . . . But you may Depend upon it they'l come to a good Market here." Employing captive Africans in the collection of plants with healing properties represented another way Chandos sought to bring greater profits into the Royal African Company's coffers through the extraction of African natural knowledge and labor.[27]

Chandos and his allies on the Court of Assistants requested specimens of West African natural commodities in sufficient quantities "to make Tryalls & Experiments with" in the metropole.[28] Collections such as those transported on the *Clarendon* allowed company officials and metropolitan naturalists to undertake further tests on specimens, especially those that West Africans used as drugs, dyes, perfumes, and food stuffs. As the duke reminded the agent at Whydah, "You'll remember to send over Samples of Gums & Roots in Quantities large enough to make experiments of."[29] In addition to these general instructions, Chandos added his own requests for particular plants or minerals upon which he wished to conduct experiments. For example, Chandos followed a general reminder to acquire gums and roots with a request that the agent at Whydah send over by the next company ship "a Jarr (well stopt)" of locally produced indigo. The duke explained that he "intend[ed] to make some Experiments here of it."[30] Chandos instructed agents in West Africa to send barrels of earth that would be tested in London for the presence of precious minerals. He even had the ballast stones on some company vessels pounded to powder and analyzed but reported he "found nothing worth the trouble in them."[31] Two years later the duke declared that he had it on good authority that the valuable medicament cinchona (the natural source of quinine indigenous to the Andes) was actually the bark of the mangrove tree that grew plentifully along the West African coast. He therefore instructed agents at Cape Coast Castle to obtain twenty pounds of bark from each type of mangrove and send it to London for trials. Empirical tests in London confirmed the Cape Coast agents' assertion that Chandos had been misinformed.[32]

To assist in the search for drugs, dyes, and other botanical commodities, Chandos pledged to recruit botanists, apothecaries, and other individuals well versed in natural history. He encouraged slaving agents to be diligent in their survey of botanical commodities with the promise that "if you can come to discover any Drugs or Gums, of which I am persuaded there must be great quantities," then Chandos would "make it my business to get as able Botanist or two to go over."[33] The duke turned to his friends in the Royal Society of London, especially the physician and naturalist Hans Sloane, for help finding an "able Botanist or two." Chandos hoped that Sloane could recommend a naturalist willing to spend six months or a year investigating the natural commodities in the vicinity of the company's slaving forts. The duke explained that these regions "abound[ed] with a vast variety of Plants & spices, whose natures & virtues are almost wholly unknown." He predicted that the generous terms the company planned to offer a botanist would "in all probability... prove the making [of] his fortune." If a dedicated naturalist could not be found, Chandos suggested that perhaps the geographies of the British slave trade would enable one to make a brief stop in West Africa on the way to the Caribbean. In 1721 the duke learned that a naturalist would soon be traveling to the West Indies at Sloane's recommendation. Chandos asked Sloane to "prevail upon" the naturalist to first stop in "this part of Africa" to survey the region's natural resources before continuing on to the Caribbean. The Royal African Company could with "great ease carry him to any part of the West Indies in one of their Negro Ships," Chandos declared.[34]

Despite Chandos's many attempts to recruit a botanist, he appears to have never succeeded. Instead, he focused on hiring surgeons and apothecaries for the company's trading factories who were well versed in natural knowledge. Paul Moze, for example, was sent in 1724 as surgeon to the company's factory at Sierra Leone. Chandos recommended Moze to the protection of the factory's commanding officer by noting that Moze was reportedly not only a skilled surgeon but also a good botanist. Moze therefore promised to be of great use to the African Company "by the Discoveries he may be able to make of divers kind of Druggs, & Roots proper for Physick & which undoubtedly must grow in great Quantities in the Wood about you."[35]

The duke and his allies also assumed that great quantities of staple commodities could be produced by the Royal African Company in West Africa. They revived the idea of establishing plantations near the company's factories, which an earlier generation of company leadership had experimented with during the late seventeenth century. Rather than produce crops such as indigo and sugar in the Americas using enslaved labor transported across the Atlantic, Royal African Company officials attempted to cultivate these products in African plantations. For example, in the 1680s the Royal African Company sent seeds, equipment, enslaved people, and experienced personnel from the West Indies to establish plantations of indigo along the Upper Guinea coast. Similarly, the company attempted to produce potash in Sierra Leone in 1698 and to grow sugar, ginger, indigo, and cotton along the Gold Coast beginning in 1703. Ultimately, the company's African plantations failed for a variety of reasons, including undercapitalization, mismanagement, and insecurity of possession at a time when Europeans projected little political or military power in West Africa.[36]

The African Company's efforts to establish plantations in the 1720s involved many of the same commodities of interest to their predecessors decades earlier. Chandos asked agents about the possibility of producing cotton, indigo, turpentine, potash, and rum from locally grown sugarcane.[37] In 1721, for example, Chandos told James Phipps, commander at Cape Coast Castle, that he and other assistants were certain of the "great advantages" that would result from Phipps's efforts cultivating cotton, indigo, rum, and ginger. Since these products would be grown in the company's "own territory," the duke anticipated that they could be produced with little expense. He also promised to send Phipps "hands skill'd in the making & dressing" of plantation commodities from the West Indies.[38] Despite such promises, company agents often cited a shortage of labor as a major impediment to implementing such plans. Agents at Cape Coast Castle, for example, explained that while they could secure enough land near the castle "for the planting of Sugar Cane Indigo Cotton Corn &c," the "industrious husbandmen" necessary for undertaking such agricultural projects were "wanting." They noted that the castle had few white employees and the "Common people" among them, such as soldiers, "for the Generallity are of so lazie a disposition that they will be at the expense of a Servant to boil the pot." Given that free Africans had little incentive to engage in such labor, the agents concluded they would need the company to provide enslaved Africans dedicated to the project before they could begin to produce indigo, cotton, sugarcane, or corn.[39]

In letters to Chandos and other members of the Court of Assistants, African Company agents took pains to demonstrate that they had followed instructions to survey, locate, extract, and introduce natural commodities into the regions surrounding the company's West African slaving factories. Their few successes were often overshadowed by their many failures. For example, agents at Sierra Leone and Whydah reported that the prickly pear cactus growing in the factory's garden contained no cochineal insects and that the cinnamon trees introduced by the slave ship captain Samuel Barlow struggled to survive.[40] Richard Hull, the company's chief agent at Gambia, advised in 1724 that "the country up the River [was] barren as to Dying Woods or Medicinal Plants. Nor any such thing as Salt Petre." Hull also reported that locusts had destroyed their cotton and indigo plants and apologized for the excessive cost of their attempts to produce indigo.[41] As agents Plunkett and Archbold at Sierra Leone reassured the Court of Assistants in 1722, they had "been very earnest to make new discoveries" yet rarely had much to show for their efforts.[42]

The Royal African Company's efforts to establish plantations at their slaving factories in the early 1720s were no more successful than its earlier attempts. Company agents did have some success collecting botanical and mineral specimens from West Africa. With varying levels of diligence, African Company agents bioprospected in the vicinity of company slaving factories in order to identify potential natural commodities. Their instructions to do so were often framed by claims that the "natures & virtues" of the African natural world were "almost wholly unknown."[43] While such statements may have characterized the state of British natural historical knowledge about West Africa, the virtues of African *naturalia* were well known within the natural and medical knowledge systems of the peoples along the West African coast.

APPROPRIATING AFRICAN KNOWLEDGES

In 1721 Chandos turned to his friend the physician and naturalist Hans Sloane to instruct Royal African Company slaving agents about how best to bioprospect in West Africa. He asked the naturalist to draw up "some rules whereby the knowledge of the Natives & usefulness of Plants or drugs or woods might the better [be] known."[44] Chandos and his allies on the Court of Assistants relied on the assistance of metropolitan naturalists such as Sloane to provide slaving agents guidelines for extracting African specimens and knowledges. They also turned to metropolitan naturalists to assess the value of the specimens slaving agents sent in response. Royal African Company leaders instructed their agents to investigate how West Africans derived dyes, perfumes, and other natural commodities from local flora, fauna, and mineral resources. Above all, they instructed their slaving agents to observe the medical practices of local Africans and to gather samples of the plants they used as drugs. Chandos and his allies frequently urged the company's slaving agents to collect natural knowledges in order to find new ways for the company to become profitable.

In response to such instructions, the African Company's agent at Whydah, Ambrose Baldwyn, sent the Court of Assistants a collection of plants and roots known locally for their healing powers. The collection included a balsam that Baldwyn reported was known among the Whydah as "Bonibos Tarr" and used with great success to cure gangrene and gout. Baldwyn's wife, who had recently returned from the Bight of Benin, testified before company officials in London that she had successfully used the balsam to treat a wound on her husband's arm. Chandos gave samples of Bonibos Tarr to British physicians of his acquaintance to test on their patients. Based on their "good Success" using "it upon Rheumatisms," Chandos asked Baldwyn to send as much of it as he could procure. The balsam was also part of a small collection of plants and drugs from Whydah that Chandos asked Sloane to evaluate. Sloane's analysis of the Whydah specimens was largely inconclusive because so many plants had spoiled during transit and, of those that survived, the small quantity sent meant that "no Sufficient experiment cou'd be made to Discover their Virtues." Sloane drew what conclusions he could about the balsam, which Chandos sent to Baldwyn along with Sloane's directions for better preserving and packing specimens in the future. The duke also passed along a reminder from Sloane that plants needed to be gathered in their proper season when the fruit and flower were available and that the entire plant, root and all, should be sent to London.[45]

The *Clarendon* collection that opened this chapter similarly reflected the African Company's efforts to extract West African natural knowledge, especially medical knowledge. All twenty-one of the plants in the *Clarendon* collection were used as remedies by West Africans, likely the Fante, along the Gold Coast. The collection included a plant "outwardly applied for the pain of the head or pluritic pains," another "to cure sores or boils or any breakings out in the flesh," and one for relief from "gripes or pains in the belly." It included specimens of two different plants used by women after childbirth as well as plants used for sore throats, swelled testicles, and rheumatic pains. Detailed descriptions of how to prepare the medicaments derived from the plants, where the plants were found, and how else they might be used accompanied the specimens on their journey to London on the slave ship *Clarendon*.[46]

Although the specimens were gathered at the instruction of the Royal African Company, the slaving company's own records tell us nothing about the *Clarendon* collection beyond that it included some drugs. Instead, descriptions of the *Clarendon* collection survive in the manuscript catalog to Sloane's museum. At some point, Chandos gave twenty-one plants transported on the *Clarendon* to Sloane. Most likely, Chandos asked Sloane to examine the plants in the *Clarendon* collection on behalf of the Royal African Company, similar to his request regarding the balsam sent by Baldwyn. Sloane recorded the new botanical additions to his museum by adding each specimen to his *Vegetable and Vegetable Substances* catalog. The catalog lists the more than 12,000 specimens that originally composed this portion of Sloane's museum. Most entries in the catalog are quite brief, describing the specimen in a phrase or two and perhaps noting where the specimen was collected or from whom Sloane received it. Typical in this regard is entry 7,634, which simply states, "A pentangular fruit in the form of a calabash. From Guinea by Mr. Staphorst."[47] The entries for the twenty-one *Clarendon* plants, by comparison, are quite long. For example, Sloane's entry for *embasnobah* reads:

8124. Is a root called Embasnobah it is a small bush growing in the nature of a Cotton vine & is full of prickles & produceth a sort of a bean with a round seed. This root is esteemed by the natives very good against the dry gripes taking a small quantity of the bark of this root jam'd fine upon a stone putting nine or ten grains of Malagetta, being thus prepared, and take the quantity of a Spoonfull with water gives present ease, the root its self chewed in the mouth and likewise very beneficiall in the gripes. The Embasnobah is procured in these parts near the waterside bushes & not in the Upland Country. From the Duke of Chandos from Guinea.[48]

The *embasnobah*'s catalog entry indicated that the plant was a small, thorny bush most commonly found near the coast. It described two distinct medicaments prepared from the plant that Africans along the Gold Coast used to treat stomach ailments. The first was a paste of the root's skin pounded with malagueta pepper and dissolved in water while the second was the root itself, which was simply chewed. The length of these entries, combined with their transcription in Sloane's own handwriting, suggest the value Sloane accorded to Fante natural knowledge.

Despite the unusual level of detail preserved in Sloane's catalogs, there is still much that we do not know about the collection transported on the *Clarendon* and the individuals whose knowledge and labor it represented. The commander of Cape Coast Castle, Captain General Tinker, received Chandos's thanks for the collection, but it is very unlikely that he selected, gathered, and preserved the specimens. This physical and intellectual labor was probably performed by Fante people who controlled the Gold Coast. The specimen labels themselves may have involved one or more additional individuals. Botanical descriptions in the Sloane vegetable substances catalog suggest that the person who wrote the labels was familiar with European medical practices and natural historical knowledge systems. But, fundamentally, the true authors of the medical and natural knowledges embodied in the *Clarendon* collection are unnamed West Africans.[49]

West African medical and natural knowledges such as that embodied in the *Clarendon* collection were dynamic, cosmopolitan, and rooted in their local social, spiritual, and political contexts. Historians of early modern medicine have shown how medical ideas, healing techniques, and medical substances moved between individuals, communities, and even continents. As they did so, borrowed ideas and novel medical substances were redeployed to suit particular local circumstances. Fundamental similarities between various

European and West African medical frameworks and practices helped to facilitate such medicinal promiscuity. Throughout the early modern world, medical practice on the ground involved more collaboration, borrowing, and improvisation than learned texts might suggest. Efforts like those of Chandos and his allies within the Royal African Company's Court of Assistants sought to abstract Gold Coast natural knowledge and medicinal substances from their cultural and spiritual frameworks in order to commodify them. As Pablo Gómez reminds us, the ability of drugs from West Africa, Asia, or the Americas to become profitable commodities within European pharmacopeias was rarely due to what in modern terms we would call the medical substances' inherent pharmacological properties. Instead, it depended upon the ways the substances were employed within cultural, spiritual, and performative contexts by the healers utilizing them. Many of the medical substances employed in both early modern European and West African healing traditions may seem implausible to modern eyes. Yet contemporaries understood them to be efficacious and were often willing to pay exorbitant prices to obtain them.[50]

In early eighteenth-century Britain, foreign medicinal plants that could be used as specifics represented a particularly promising commercial opportunity. Consequently, Chandos and his allies repeatedly instructed Royal African Company slaving agents to search for new specifics. For example, in 1721 Chandos ordered James Phipps, captain general at Cape Coast Castle, "to enquire into the Physick the Natives use for curing distempers & if they have any Specificks to find out of what they are made." Phipps, Chandos continued, should send samples of the drugs used to treat ailments such as fevers and fluxes, along "with an Account of their Nature & how used & in what quantities."[51] Chandos sought information about Fante medicaments, including whether they included any "Specificks."

"Specifics" refers to remedies used in a targeted manner to treat a particular ailment. They were premised on an alternative medical framework to the long-dominant Galenic medical theory, which asserted that illnesses resulted from an imbalance in a body's humors, were influenced by their environment, and needed to be corrected through a highly individualized regime to restore balance. Competing understandings of illness that had long existed but grew in popularity beginning in the late seventeenth century understood diseases as discrete entities that could be treated through the application of the correct specific. Specifics were understood as universal therapeutics that should work on any body afflicted with the corresponding disease, regardless of the humoral constitution of that body and irrespective of its environment. Specifics could be moved great distances and still be expected to be efficacious.[52]

Specifics were therefore particularly well suited to the search for and commodification of exotic drugs through the routes of global commerce and colonialism. Expanding British colonialism brought more Britons to unfamiliar and deadly disease environments and far from the medical marketplaces they had once relied upon. It created a need for standardized medicines that would work anywhere, on any body. The labor needs of an expanding British Empire, especially needs associated with the plantation complex, the British military, and the East India Company, encouraged the dramatic expansion in the market for specifics over the course of the eighteenth century.[53] The most popular specifics sold in Europe in the eighteenth century were imported from abroad. Further, European physicians and naturalists believed that the identification of new specifics was most likely to come through the appropriation of medical knowledge from non-European peoples. As French natural philosopher Pierre-Louis Moreau de Maupertuis declared in 1752, "it is quite by accident and only from savage nations that we owe our knowledge of specifics; we owe not one to the science of physicians."[54] Foremost among the drugs Maupertuis likely had in mind was cinchona, the natural source of quinine, which was highly effective against intermittent fevers. Jesuit missionaries had learned about the healing properties of the cinchona tree's bark (or "Jesuit's bark") from the Indigenous people of Loja in modern Ecuador. Chandos may also have been thinking of cinchona when he urged the commander of Cape Coast Castle to pay particular attention to specifics for fevers and fluxes.[55]

The Royal African Company asked its slaving agents to search for specifics and other medical substances as part of its own search for new ways to profit through trade with West Africa. The African Company in the 1720s instructed agents to collect natural knowledges as well as natural objects in the vicinity of its slaving factories in West Africa. It then enlisted metropolitan naturalists and physicians such as Sloane to test the specimens sent by its slaving agents in hopes of identifying new drugs and other natural commodities. In some ways this paralleled the place of non-European natural knowledges in the Americas. Naturalists in eighteenth-century British plantation societies, for example, keenly investigated the natural and medical knowledges of enslaved Africans and Native Americans, yet maintained that what they collected did not constitute natural knowledge. Instead, colonial naturalists portrayed such knowledges as merely the raw materials out of which they fashioned new knowledge through additional experimentation and study.[56] The failures of the Royal African Company's efforts to commoditize Gold Coast medical knowledge are in some ways the most interesting part of the story. The Royal African Company's inability to identify West African drugs in which the company might trade is part of a larger pattern identified by historian Benjamin Breen. Drugs from West and West Central Africa are notably rare in early modern European pharmacopeias compared to those from other regions of the world. This was despite long-standing commercial ties between Europe and sub-Saharan Africa and the importance accorded to healers of African descent in the diaspora. Only in the nineteenth century did African medicaments become an important part of the global drug trade.[57]

AN INFRASTRUCTURE FOR ENSLAVING AND COLLECTING

In January 1721 the Court of Assistants issued a new standing order to the Royal African Company's agents in West Africa. When enslaved Africans arrived at the company's coastal forts, agents were instructed to interview them "concerning the Country from whence they came" before selling them into the transatlantic slave trade. The court instructed agents to "very strickly examin[e]" captive Africans regarding the geography, natural commodities, political structure, and commercial potential of their homelands. The assistants told the agent at Cape Coast Castle to ask about the "situation, Rivers, Soils, Commodities, . . . Government, [and] its distance from Cape Coast." The company was particularly interested in the commercial potential of inland regions. Agents were therefore instructed to inquire into both commodities grown in captives' homelands as well as those imported into it. The court was keen to know whether a trade in gold, silver, nutmegs, cloves, or drugs might be possible and "by what means" captives thought such a trade "might be brought about." If any captives seemed likely to open such a trade with the company, the court empowered agents to free them.[58] Chandos instructed company agents to record what they learned from their interviews and send copies of these records to the Court of Assistants each year. He predicted that in time such "frequent Examinations" would enable the company "to come to a better Knowledge of the Inland parts of Africa, than we have been hitherto able to attain."[59]

The Royal African Company's standing order to interview captive Africans arriving at its coastal slaving forts explicitly relied upon the infrastructure of the slave trade as a means of gathering natural historical, commercial, geographic, and political knowledge of "Inland parts of Africa" about which Europeans knew relatively little. The geographic reach of the slave trade within West and West Central Africa was wide. Enslaved Africans might be forced to walk hundreds of miles from where they were originally kidnapped or enslaved to the coast where they were eventually sold to a European slave trader. Consequently, many captives arriving at coastal slaving forts like Cape Coast Castle hailed from inland regions of Africa where Europeans had never visited. When explaining the new standing order to a correspondent, Chandos highlighted that captives often traveled for months before they reached the company's slaving factories. He thereby acknowledged how the immense sweep of the brutal trade increased its potential value as a vector of natural knowledge collection.

More broadly, Chandos and his allies on the Court of Assistants sought to use the infrastructure of enslaving to gather information about how better to exploit West Africa. They collected knowledge about African natural commodities and potential commercial opportunities by means of the British slave trade's communication networks, trading routes, and personnel. They instructed company employees to transport seeds, specimens, botanical samples, and written descriptions of West African *naturalia* between their slaving factories in the Atlantic World and their headquarters in London. The Court of Assistants not only understood the strategic importance of natural history for promoting commerce, but they also suggested ways that the commercial infrastructure of the slave trade could be harnessed to collect natural knowledge. The Court of Assistants' instruction to interview captives at the company's slaving factories was thus part of the company's efforts to exploit the extensive reach of slaving networks in order to collect natural and commercial knowledge.

To facilitate "frequent Examinations" of captive Africans arriving at coastal slaving factors—and to help to bridge linguistic barriers—Royal African Company officials instructed their agents to show captive Africans samples of the natural commodities the company sought.[60] While slaving agents were likely to have near to hand natural commodities already exchanged in the West African trade such as gold dust and ivory, they were unlikely to have easy access to botanical commodities in which the company was also interested. The Court of Assistants therefore sent chests containing samples of drugs, dyes, and other natural commodities to each of their West African factories in the 1720s. Slaving agents were instructed to show captive Africans the plant specimens contained in each chest. Chandos explained to Sloane that the chests of specimens would be used "in order to make inquiry amongst the Natives & Negroes brought down (as some are Three months journy) to be sold, whether their country produces any such." Boxes of botanical samples were intended as tools for extracting African natural knowledge.⁶¹

The African Company's slaving fort at Gambia received its box of botanical samples in 1722. The Court of Assistants explained that the samples of drugs and spices were intended to guide the search for drugs and other useful plants. For that reason, "we desire particular care may be taken to preserve them." The slaving agents at Gambia were instructed to show every species of plant contained in the box of specimens to enslaved Africans arriving at the trading fort. Slaving agents were expected to conduct "a full examination" into "whether any such grow in the part from whence they came." Company officials believed African merchants from whom they purchased captives were also potential sources of information about useful local plants. Therefore, the company's agents were instructed to show the botanical samples to local African merchants and rulers "with whom you are in alliance, & prevail with them to cause Search to be made in their Woods & proper places to see if they can find" any of the plants contained in the box.⁶²

The contents of the boxes of botanical samples were in part determined by Sloane. In December 1721 Francis Lynn, secretary to the Royal African Company, called upon Sloane to show him the list of botanical samples that the company planned to send to its slaving factories in West Africa. In the accompanying letter, Chandos "intreat[ed] the Favor" that Sloane would "cast your eyes over" the list and determine whether "there is any in this Catalogue too common to be worth the inquiring after, or if there are any other which was omitted, & which you think proper to be included." A few weeks later, the Royal African Company formally thanked Sloane for his "amendments" to their list of samples and for the trouble he had been at to assist them. The historical record gives no hints about what amendments Sloane offered or the exact contents of the boxes of botanical samples.⁶³

Royal African Company leaders relied upon the knowledge of the company's returning slave ship captains, ship surgeons, and agents in the West Indies to inform their efforts to identify, introduce, and trade in natural commodities. Former Royal African Company employees were among Chandos's most important sources of information about the commercial potential of the regions where the company traded. Chandos and the African Company's secretary, Lynn, interviewed slaving agents, ship captains, and surgeons after they returned to Britain. From Seth Grosvenor, a former agent at Cape Coast Castle, they learned that the castle's garden contained cinnamon trees and prickly pear cacti. Chandos therefore directed Grosvenor's successor to report whether the cinnamon trees were still alive and search the prickly pear cacti for cochineal beetles, the source of a valuable red dye imported from Spanish America.[64] Capt. Samuel Barlow, who commanded three slaving voyages for the Royal African Company in the early eighteenth century, reported that large indigo trees grew in Whydah and that Cape Coast abounded with the same exotic woods exported from Sierra Leone.[65] Chandos consulted other slave ship captains about the possibility of cotton and ginger production, about when Guinea grain at the Bight of Benin ripened, and about the amount of balsam that could be exported from Whydah. Since most slave ship captains traded in natural commodities as well as human captives, veterans of the trade like Barlow would have been invaluable informants regarding African natural commodities.[66]

Chandos similarly enlisted the assistance of Royal African Company agents in the West Indies. For example, the company's agent in Jamaica, Dr. Stewart, provided his counterpart in Gambia with samples and preparation instructions for a medicinal root known in Jamaica as *contra yerba*. According to Stewart, captive Africans arriving in Jamaica "know it immediately upon seeing it" and reported that it was native to West Africa where it "grows in great plenty on the Sea Side." Chandos observed that the plant was known by different names "according to the Language of the Country" where it grows, but that most captive Africans "know it by the Name of Assintee i.e. Bitter Grass." *Assintee*'s presence on both sides of the Atlantic may itself reflect the impact the slave trade had on the ecosystems of the Atlantic basin. Judith Carney and Richard Rosomoff have shown how "the African diaspora was one of plants as well as peoples." The forced migration of millions of Africans led to the intentional and unintentional transfer of African plants and animals. The sophisticated and specialized botanical and agricultural knowledges captive Africans carried with them enabled the acclimatization and cultivation of African plants in plantation societies in the Americas.[67]

Chandos instructed slaving agents in Gambia and the Gold Coast to search for the *assintee* and to experiment with cultivating it near the company's factories. The duke declared that the drug was "very wholesome" to mix into the food given to captive Africans during the middle passage and would "prevent their falling into those distempers which a long voyage subjects them to." Chandos did not need to explain to his correspondent that reducing sickness and mortality on board the company's slave ships would improve its profits. The duke identified a second way to profit by means of African knowledge about the *assintee*. Chandos concluded based on samples of the plant sent by slaving agents at Cape Coast Castle that the plant known as *assintee* along the Gold Coast was not the same species as the Jamaican *contra yerba*. But as the two had similar healing properties, the duke believed "'tis very probable our Druggists will be pleased to see it at Market." He therefore ordered the company's agents in West Africa to send 200 or 300 pounds of the *assintee* root. Chandos sought to doubly profit from the extraction of African knowledge about the *assintee*, first as a tool employed in the commodification of enslaved Africans and second as a commodity in its own right to be sold in British apothecary shops.[68]

The commercial infrastructure of the slave trade enabled the African Company to enlist its employees in the West Indies in support of other efforts to improve its profits. Chandos instructed the Jamaican agent Stewart to supply the company's factory at Sierra Leone with the supplies and skilled laborers necessary to establish a plantation of indigo there. In December 1723 Chandos reassured Stewart that the Court of Assistants "are all pleased" with Stewart's "Endeavours to procure some Negroes versed in the manner of making" indigo from Jamaica. Once the agent had purchased at least ten such enslaved Africans, Stewart was to make arrangements to send them from Jamaica to Sierra Leone. Chandos instructed Stewart to put the enslaved Africans on a sloop bound for Sierra Leone, laden with rum and "a good Quantity of Indigo Seeds." The entire cargo was to be consigned to Mr. Archbold, the chief agent at Sierra Leone. Archbold, Chandos explained, would be instructed to send the sloop back to Jamaica with 60 to 100 enslaved Africans on board.[69]

The company's orders to Stewart reflected how the Court of Assistants employed the broader commercial resources of the company in its search for natural commodities that could lead to new branches of trade. The African Company sought to obtain natural and commercial knowledge from captive Africans themselves, as well as from the slaving agents and mariners employed by the company. Chandos and his allies among the assistants relied upon the commercial infrastructure of the slave trade to facilitate their search for natural knowledge that could make the Royal African Company profitable again. The company transported seeds, specimens, supplies, and skilled laborers on its ships and instructed employees throughout the Atlantic World to assist in their efforts to diversify the company's African trade. African Company agents integrated these cargoes of seeds and specimens into the business of enslaving.

CONCLUSION

The Royal African Company's efforts to identify new ways to profit through trade with West Africa relied on the collection of natural historical objects and knowledges. It reflected the faith placed in natural history to benefit commerce and colonialism in the eighteenth century. The company's efforts hinged on the belief that the appropriation of African natural knowledges and bioprospecting for natural commodities might enable the company's leadership to identify new ways for the company to profit. The infrastructure of enslaving, including the Royal African Company's slaving factories, vessels, agents, and ship captains, proved central to its efforts to survey West Africa's botanical and mineral resources. Ultimately, the company's efforts to revive its fortunes through the collection of natural objects and knowledges proved unsuccessful.

The natural historical specimens collected by African Company employees may have been relatively small in absolute numbers but were significant given how few West African specimens could be found in early eighteenth-century British museums. For example, the twenty-one plants gathered near Cape Coast Castle in the *Clarendon* collection represent a significant fraction of Sloane's total collection of African plant material. Victoria Pickering's definitive study of the 12,523 items in the Sloane *Vegetable and Vegetable Substances* catalog indicates that less than 4 percent of the specimens in the catalog for which geographical information is available were from Africa. The vegetable catalog lists 188 specimens from Africa, mostly from "Guinea" (West Africa) or "the Cape" (South Africa). Nearly half of these—90 of 188—were given to Sloane by Chandos. Chandos in turn almost surely acquired them by virtue of his position within the Royal African Company.[70]

Chandos's efforts to revive the finances of the Royal African Company were short-lived and unsuccessful. Although the company's fortunes looked promising in the early 1720s with a flurry of new investment and trading activity, by 1725 it was in financial difficulties again, and its days as a transoceanic trading house were largely over. Chandos withdrew from active involvement with the company by the end of 1724. The Royal African Company sent only five additional slaving voyages over the next twenty-five years. Instead, it largely subsisted on annual Parliamentary subsidies for the upkeep of the company's factories, which the company had long argued benefited all Britons engaged in the slave trade. Rumors of financial mismanagement led Parliament to end the subsidies, and the company formally dissolved in 1752.[71]

In 1713, the year after the Royal African Company fully lost its monopoly, British merchants formed a new monopolistic slaving company. The Treaty of Utrecht granted Britain exclusive rights to engage in the transatlantic slave trade to Spanish America at the conclusion of the War of Spanish Succession. The British Crown, in turn, granted these exclusive trading privileges to the newly formed South Sea Company. As we will see in the next chapter, the slave trade to Spanish America provided unique opportunities for British naturalists in the early eighteenth century to obtain specimens and to study the natural world in a region officially off limits to them.

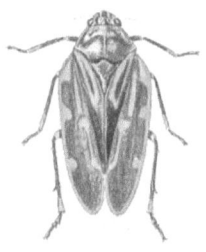

3.

The *Asiento*'s Natural Historical Profits

Officially, at least, Britons were not welcome in eighteenth-century Spanish America. British naturalists interested in cataloging valuable natural commodities indigenous to Spain's colonies were perhaps even less welcome. But for twenty-six years during the early eighteenth century, a small group of British surgeons, ship captains, merchants, and mariners who were engaged in the slave trade were allowed to visit and even reside in select Spanish American port cities with the blessing of the Spanish Crown. A few of these British slave traders used their access to Spanish America to acquire natural historical collections. The *asiento* slave trade created an opening that British naturalists eagerly sought to exploit in order to study the flora and fauna of a region rumored to abound in drugs, dyes, minerals, and other natural wonders.

John Burnet was among the small group of British slave traders resident in Spanish America during the early eighteenth century because of the *asiento*. He spent his days treating the ill and the dying among captive Africans at the British South Sea Company's slaving factory, or trading fort, in Portobelo on the Isthmus of Panama and, later, in Cartagena de Indias in what is now Columbia. As a factory surgeon, Burnet "Administer[ed] Surgery & Phisick out of the Company's Medicines to the Factory whenever there shall be occasion." He was instructed to visit the company's enslaved Africans "as often every day as occasion may require" and to take particular care of the sick among them.[1] During the years that Burnet worked as a factory surgeon he also quietly collected scores of natural historical specimens on the side. He employed the commercial routes of the *asiento* slave trade to transport flora, fauna, minerals, and natural observations to Britain.

In an era when few foreigners were officially allowed to visit Spanish America, Burnet spent more than a decade residing there. He and other employees of the South Sea Company were among the Britons with access to investigate Spanish America's flora and fauna firsthand and surreptitiously to collect natural historical specimens. South Sea Company collectors relied on the personnel and infrastructure of the slave trade to acquire specimens and to transport them to Europe. Although only a handful of South Sea Company servants undertook such investigations, their efforts uniquely shaped natural history. The seeds, specimens, and observations they gathered enriched the British herbariums, botanic gardens, and museums that were essential to the work of early modern natural history. South Sea Company employees provided access to Spanish American specimens and natural knowledges otherwise difficult for British naturalists to obtain. Their collections of objects and knowledges represent the natural historical profits derived from the British slave trade to Spanish America.

The routes of the eighteenth-century British slave trade did not correspond with the outlines of the British Empire. Slaving regularly brought Britons beyond the boundaries of British territories. More broadly, recent scholarship on the early modern Atlantic World reveals "an interconnected hemispheric system," in which the histories and economies of maritime empires were deeply entangled. Far from metropolitan centers yet close to spaces claimed by rival powers, economic opportunity in the Americas often inspired interimperial trade that violated one or more empire's commercial regulations. Imperial borders could be porous, identities fluid, and allegiances improvisational. The opportunities to cross boundaries were many at the edges of empire, as Burnet's own professional trajectory reflects.[2] Burnet began his career as a British slave ship surgeon, but he ended it as a personal physician to the king of Spain. The fluidity of the early modern Atlantic World—especially along the routes of the slave trade—enabled both Burnet's professional trajectory and the addition of Spanish American specimens to British gardens and museums.

ACCESSING SPANISH AMERICA

John Burnet's collecting efforts were a reason that men like him were not supposed to be in Spanish America in the first place. Like other European powers, Spain strove to restrict trade to within its imperial boundaries. Spanish officials knew that the value of their trade depended in part upon maintaining their monopoly on natural commodities indigenous to their empire. They understood that given half a chance, their imperial rivals would smuggle the natural sources of Spanish American dyes and drugs into their own territories. Consequently, the Spanish Crown forbade the entry of foreigners into Spanish America and attempted to guard knowledge about natural commodities indigenous to their territories. Spain's policies of secrecy and exclusion were far from absolute. People, natural objects, commodities, and knowledges circulated in defiance of official prescription.[3] Yet the perception among British naturalists maintained that relatively little was known about the natural world of Spanish American territories, a state of affairs that many British naturalists hoped the *asiento* would change.

The stakes for doing so were high. Spanish America was home to some of the most valuable natural commodities known to early modern Europe. These included cinchona, the antifebrile indigenous to the Andes that contains the natural source of quinine, and cochineal, a brilliant red dye that was more valuable by weight than silver. Spain's policies of secrecy and exclusion of foreigners largely worked for more than 200 years, leaving naturalists in other parts of Europe ignorant about much of the natural history of Spanish America. As late as 1734 European naturalists still debated the basic classification of cochineal; was it an animal, a vegetable, or a mineral? British naturalists were confident that an environment home to natural treasures such as cochineal and cinchona must surely contain others.[4]

Spain's policies of secrecy and exclusion intensified British desire for knowledge about Spanish America. The tendency for works in Spain to circulate in manuscript rather than in print meant that there was a limited pool of imported Spanish books for British printers to translate and reprint. The foreign accounts of Spanish America often described coastal areas and relied on rumors and hearsay for information about lands further inland. Travel literature about the region typically depicted the area as resplendent with valuable minerals and other natural commodities ripe for the taking. For many Britons, it reinforced the perception that vast profits could be made if they could gain a foothold in Spanish America.[5]

British naturalists had also long been eager to learn more about Spanish America. The second volume of the Royal Society of London's *Philosophical Transactions* (1667) included a series of questions that the society's fellows hoped travelers could answer about the region. These inquiries sought, in particular, to determine if the more fantastical claims made in travel literature would stand up to eyewitness inspection. In a classic articulation of the Baconian ideals upon which the society was founded, the article's introduction explained that "'tis altogether necessary, to have confirmations of the truth of these things from several hands, before they be relyed on." The article asked, for example, whether in Panama "Toads are presently produced, by throwing a kind of Moorish Water found there, upon the Floors of their Houses" as the Dutch author Jan Huyghen van Linschoten had reported. The society's interest in Spanish America also led the editor of the *Philosophical Transactions* to include reviews of travel narratives about the region among the journal's many descriptions of natural wonders, novel experiments, and other advances in natural knowledge. Naturalists, merchants, and imperial officials shared the conviction that Britain could only benefit if her subjects had greater access to Spanish territories.[6]

The slave trade represented an important commercial entrée into Spanish territories. Like colonists throughout the Atlantic World, Spanish colonials desired enslaved Africans to work in their fields, homes, textile mills, and mines. Recent research has shown that Spanish America was second only to Brazil in terms of the total number of enslaved Africans trafficked to the region. Over 2 million captive Africans disembarked in Spanish America between 1505 and 1867. Yet less than half of these captive Africans were transported on Spanish vessels. During the first half of the eighteenth century, Spain's direct participation in the transatlantic slave trade was particularly limited. Nearly every enslaved African brought to Spanish America between 1701 and 1760 was transported on a slave ship belonging to another imperial power. Some of these foreign vessels participated in the contraband slave trade to Spanish America. Yet it was the sanctioned slave trade to Spanish territories that British naturalists believed promised the greatest access to the region. In order to supply its colonies with enslaved labor, the Spanish Crown in the late seventeenth and early eighteenth centuries negotiated a series of long-term contracts for foreign traders to deliver a set number of enslaved Africans to its territories over an agreed-upon period of time. The Asiento de Negros, or *asiento*, offered its holder a monopoly on the legal slave trade to Spanish America.[7]

An illicit British slave trade to Spanish America had flourished since the mid-seventeenth century. However, the legal British slave trade to Spanish territories only lasted from 1713 to 1739, when the British held the *asiento* as a result of its victory in the War of Spanish Succession. The Treaty of Utrecht awarded the British Crown the *asiento* in 1713. Queen Anne consigned her rights to the British South Sea Company, which was set up for this purpose. In return for supplying 4,800 enslaved Africans each year for thirty years, the British South Sea Company was allowed to bring back its profits in goods, coin, or bullion. In addition, the company could send one ship each year containing British manufactured goods to sell in Spanish territories. The *asiento* allowed the South Sea Company to establish trading factories to house unsold enslaved Africans in Spanish American ports including Buenos Aires, Cartagena, Havana, Panama, Portobelo, Santiago de Cuba, and Veracruz. Each South Sea Company factory employed slaving agents to oversee the sale of captive Africans and a factory surgeon responsible for their health. Although the British *asiento* was supposed to last until 1743, it ended with, and partially caused, the War of Jenkins' Ear in 1739.[8]

Like those who had held the *asiento* before them, the British hoped that it might create an opening to Spanish American markets through which more than enslaved Africans would flow. The possibility that the *asiento* would cover a broader contraband trade was a source of tension between Spanish officials and the South Sea Company from the beginning of the contract. Spain worried that the company would smuggle manufactured goods, flour, and other provisions. Spain's officially sanctioned commercial system was perpetually insufficient to meet demand in Spanish America. For British investors, the possibility of contraband trading was part of the *asiento* trade's appeal. Merchants saw potential profits not necessarily in the slave trade itself but in the access to Spanish American markets and bullion that such a trade made possible. This expectation helped to set the stage for the British economic bubble that bears the South Sea Company's name.

Historians of the *asiento* have argued that the terms of the British contract were perfectly contrived to enable contraband trading by both the company and individuals employed in its service. The activities of South Sea Company employees such as Burnet demonstrate that flour and manufactured goods were not the only things being smuggled on board the company's vessels. They also used their access to Spanish territories surreptitiously to collect specimens, to record natural knowledge, and to gather seeds and seedlings of valuable plants. These efforts began with the departure of the very first British vessels engaged in the *asiento* trade.[9]

THE (NATURAL) HISTORY OF A VOYAGE

HMS *Warwick* sailed out of Plymouth Sound, bound for Buenos Aires, on February 17, 1715. The royal naval vessel, a fourth-rate ship of the line with thirty-two guns and a crew of 150 men, was under the command of Capt. Henry Partington. The British warship was on an unusual mission: to help establish a slaving fort in Spanish America. Its task was to transport the goods and personnel belonging to the new South Sea Company factory in Buenos Aires. Queen Anne declared that she intended this uncommon use of a Royal Navy vessel "as a favour and Encouragement to her Subjects" and out of recognition that it was "not possible for the [South Sea] Company to find Two Merchant Ships at hand capable of Carrying Six Hundred Tons of Goods each." In the weeks before the *Warwick* sailed, letters and accusations flew between Captain Partington and the newly appointed president of the Buenos Aires factory, Thomas Dover. The source of the conflict was the nine unauthorized passengers Dover had brought on board the *Warwick*. These passengers included William Toller, a surgeon whom Dover had appointed as the expedition's historian and naturalist—without the South Sea Company's knowledge or permission. The terms of the *asiento* limited to six the number of British subjects living at each factory; Dover's unauthorized passengers put the Buenos Aires factory well over its limit. Yet ultimately Partington was instructed to accept his unwelcome guests and proceed to Buenos Aires.[10]

As the *Warwick* completed its seven-month voyage to South America, Toller carefully recorded the natural world around him. He compiled his observations into a manuscript titled "The History of a Voyage to the River of Plate & Buenos Ayres from England." Toller's "History" represented the first natural history to result from the British assumption of the *asiento*. As a work of natural history, Toller's text is of limited interest. There is no evidence that it influenced the work of subsequent naturalists or that it advanced natural knowledge in any significant way. Yet its very existence is remarkable. Toller's "History" signifies the unquantifiable ways by which British natural history benefited from the *asiento* trade. The very fact that Dover, at his own expense and in violation of his instructions from the South Sea Company, decided that a voyage to transport a handful of merchants and trade goods warranted its own natural historian reflected the rarity of access to Spanish America. Just as merchants hoped that the *asiento* would provide unfettered access to Spanish American markets, so too naturalists expected that it would provide unparalleled opportunities to observe and collect the Spanish American natural world.

Toller began his observations as soon as the *Warwick* set sail. Each day Toller noted the weather and ship's course as well as the flora and fauna the expedition encountered. He drew the fish, dolphins, and birds he observed from the ship and speculated on their proper classification (fig. 3.1). The dolphin, for example, "is very unlike the Common Pictures of him" and quite distinct from the porpoise. Toller argued that his observations at sea led him to conclude previous authors had confused the two animals.[11] The *Warwick* reached the coast of South America four months after leaving Plymouth, enabling the natural historian to continue his explorations on solid ground. While Dover and others hunted for fresh meat, Toller explored the natural productions of Cape Castillos, in modern Uruguay. The surgeon gathered specimens of shells along the sandy beach and noted the bones of a large whale. He inventoried the plants and soil indigenous to the nearby mountains, hills, and plains. He observed the abundance of "good & Fat" wild fowl and reported sightings of deer, polecats, and armadillos, including two that were brought back to the ship as pets. Toller gathered samples of plants and hunted for birds and other small animals in order to provide models for his sketches of the local flora and fauna. As the *Warwick*'s crew struggled to find a passage through the shoals of the Río de la Plata during the next three months, Toller continued to observe the Spanish American natural world that had largely been off limits to previous Britons.[12]

Toller's investigations emphasized utilitarian natural knowledge of the region. As he declared in the opening lines of his "History," his objective was to "inform my self of every particular Necessary" that would benefit the South Sea Company.[13] He understood natural knowledge could benefit commerce as much as commerce promoted the production of natural knowledge. Above all, Toller's "History" inventoried the economic potential of the region. His study of the plants, soil, and animals of Cape Castillos, for example, led him to conclude, "The prospect of this Place is very pleasant, but I cannot incourage a settlement Here." He pointed to the region's "poor, & Hungry" soil, bad water, and lean cattle as evidence that the region was unsuitable for settlement.[14] As historian Adrian Finucane has shown, part of British enthusiasm for the *asiento* trade was the hope that it would facilitate British settlement in regions claimed by Spain.[15] Natural historical surveys such as that undertaken by Toller in Cape Castillos would help to determine which areas were worth taking and how best to exploit the resources they contained.

Toller's text also included a range of cartographic information, including detailed sailing instructions, charts indicating soundings taken by the *Warwick*, and a series of coastal profiles. "The Plan of the Bay of Castillos,"

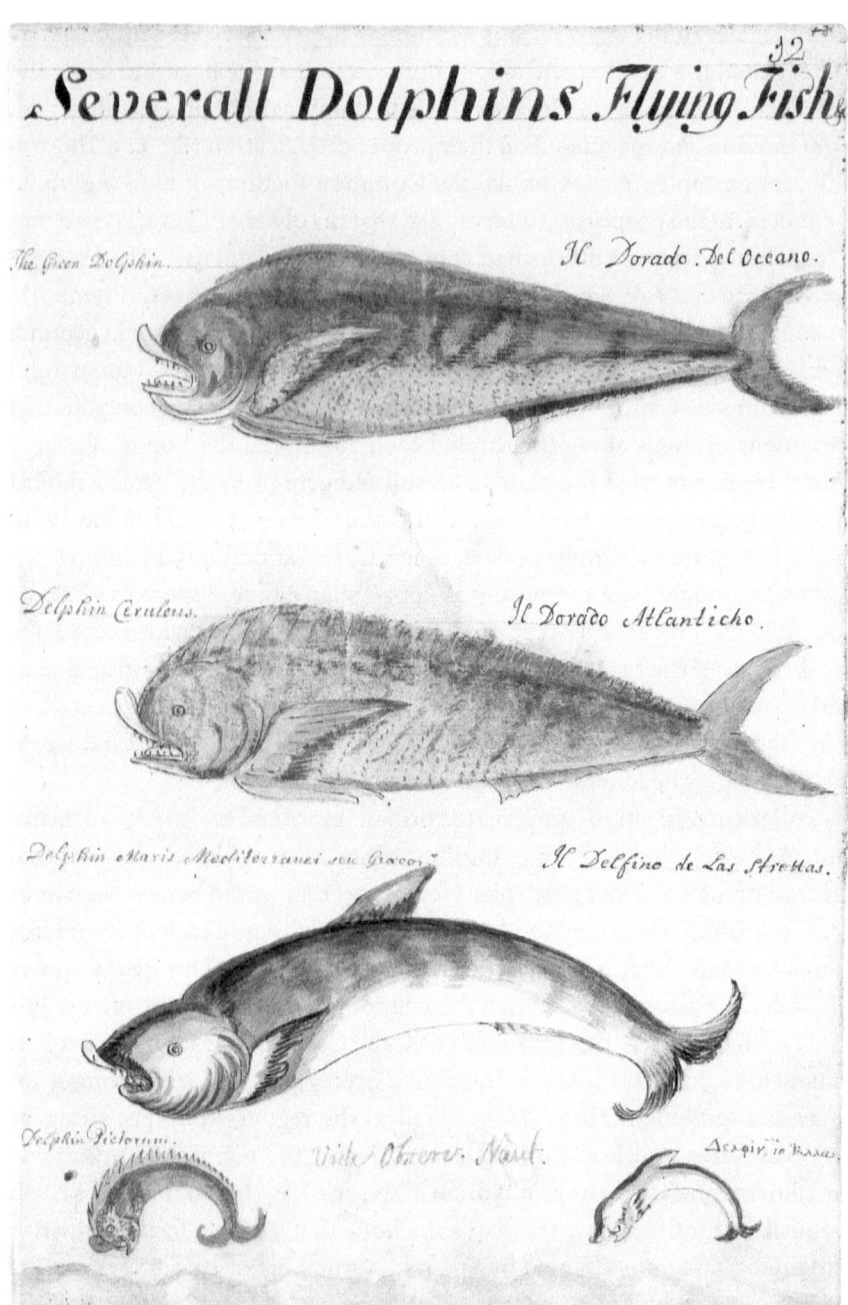

FIGURE 3.1. Dolphins and fish observed by William Toller. William Toller described and sketched fish, birds, and other animals he observed while on board the HMS *Warwick*, including these fish and dolphins. William Toller, "The History of a Voyage to the River of Plate & Buenos Aires, from England (1715)," Mss 3039, Biblioteca Nacional de España. Image taken from the holdings of the Biblioteca Nacional de España.

for example, combined nautical information with knowledge of the surrounding countryside gathered during Toller's terrestrial explorations (see fig. 3.2). It indicated the river's depth at various points, where fresh water and wood could be found, and the locations of savannas, mountains, and pasturage further inland. In a classic act of appropriation, Toller's maps commemorated the voyage of which he was a part, labeling part of the interior "Dover's Chace," another "Pas[s] de Toller," and a portion of the river "Warwick Bay."[16]

Improved cartographic knowledge of Spanish America would facilitate British commerce in times of peace and military conquest in times of war. The *Warwick*'s difficulty navigating the shallow waters of the Río de la Plata demonstrated the critical importance of detailed and accurate cartographic information. Captain Partington struggled to locate channels in the Río de la Plata deep enough for the *Warwick*. Meanwhile, the South Sea Company factors on board worried about profits lost by delay if the transatlantic slave ships they were supposed to meet in Buenos Aires had already arrived. Slave ship captains could not sell their human cargo while the company's factors remained on board the *Warwick*. Every captive African who died in port reduced the company's profits. Partially in response to this, Dover eventually sent two of the factors ahead to Buenos Aires in the company's yawl. He instructed the factors to employ a local pilot to guide the *Warwick* into port, to conduct "preliminarys with the Gov[erno]r," and "to take care of the Slaves if any Guinea Ship is arrived."[17] Detailed, accurate cartographic knowledge might have enabled the crew of the *Warwick* to navigate the Río de la Plata more efficiently. Toller claimed that his "History" provided just such natural knowledge. He declared that his text and its accompanying maps were much more accurate than the French chart the *Warwick* had relied upon. Toller predicted that his "History" would aid "all Persons who may have occasion to come on this Coast, or who are desirous to be justly informed of it."[18]

British naturalists, imperial officials, merchants, and military officers shared the desire "to be justly informed" about Spain's valuable American territories. Toller's "History" reveals that from the very first South Sea Company voyages, Britons converted the access created by the *asiento* slave trade into opportunities to collect specimens and observations about the flora, fauna, peoples, and lands over which the Spanish Crown claimed dominion. By so doing, they promoted British commercial and colonial ambitions as well as the ambitions of British naturalists who hoped the slave trade to Spanish America would profit the production of natural knowledge.

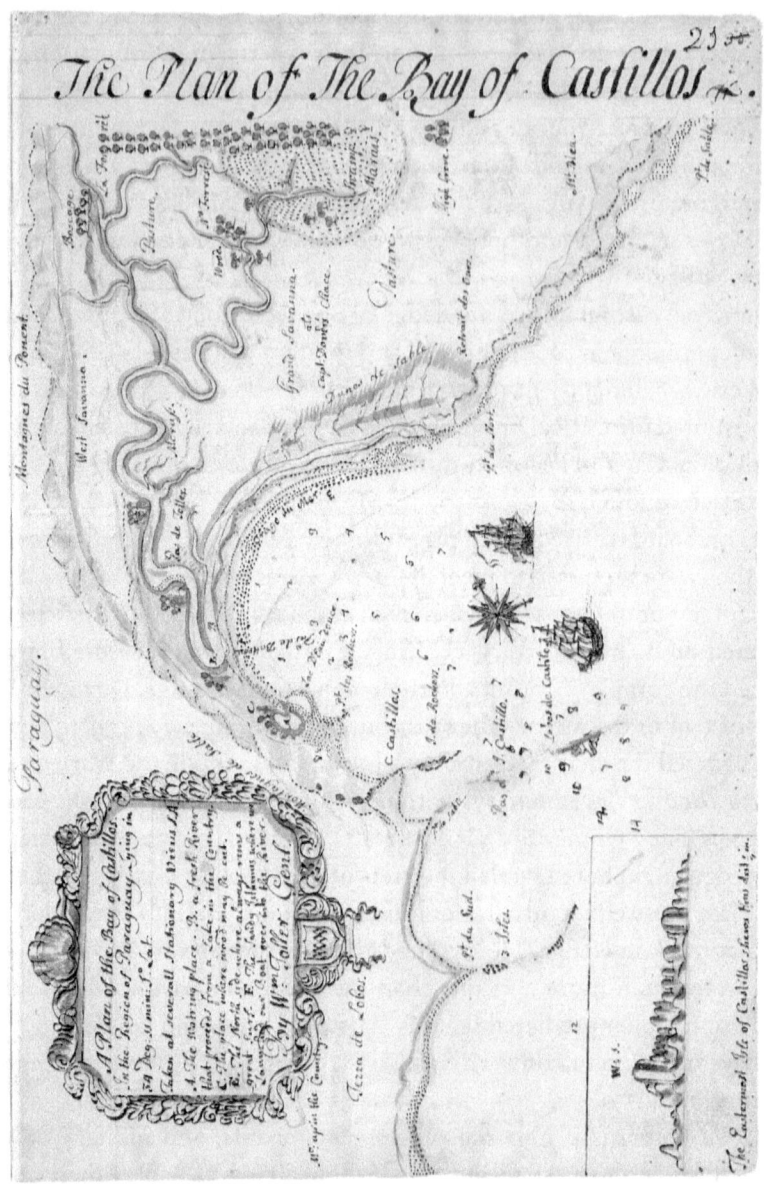

FIGURE 3.2. Map of the Bay of Castillos. William Toller compiled this plan of the Bay of Castillos, in modern Uruguay, based on his collecting expeditions and observations from HMS *Warwick*, a royal naval vessel that transported the personnel and trade goods necessary to establish the South Sea Company's slaving factory in Buenos Aires. William Toller, "The History of a Voyage to the River of Plate & Buenos Aires, from England (1715)," Mss 3039, Biblioteca Nacional de España. Image taken from the holdings of the Biblioteca Nacional de España.

COLLECTING AT THE EDGES OF EMPIRE

Ambition in large part explains the natural historical collection on board one of the slave ships awaiting the *Warwick*'s arrival in Buenos Aires in 1715. The plants, shells, fish, and shark's jaw stowed on board the slave ship *Wiltshire* belonged to ship surgeon John Burnet, who had gathered the objects during the slaving voyage to Buenos Aires. Burnet continued to acquire natural historical specimens that he gifted to British collectors over the course of the nearly fifteen years he worked for the South Sea Company. The slaving surgeon leveraged the access afforded by his residency in Spanish America as a surgeon at a slaving factory to gather rare and valuable medicaments, dyes, animals, and other natural historical specimens. Burnet's collecting efforts reflected a strategy for professional advancement as much as an interest in the natural world. He hoped that his metropolitan correspondents would repay his collecting efforts by advocating on his behalf with the South Sea Company's Court of Directors. Burnet's career as a collector and a slaving surgeon reflected the fluidity of allegiance and identity that characterized the multi-imperial Caribbean of the early eighteenth century.

Burnet acquired Spanish American flora and fauna throughout his career with the South Sea Company. He entered the company's service in 1715 after studying medicine at the University of Edinburgh. Burnet's first posting was as the *Wiltshire*'s ship surgeon. The surgeon acquired at least twenty-three specimens during the *Wiltshire*'s slaving voyage to Buenos Aires. Burnet presented this collection to the South Sea Company's Court of Directors upon his return to London. This gift, combined with a personal recommendation from Dover, secured for Burnet a more permanent post as a factory surgeon beginning in 1716. The South Sea Company initially appointed Burnet factory surgeon in Portobelo, the Atlantic port city located a three-day journey across the isthmus from Panama. When the War of the Quadruple Alliance broke out in 1718, Burnet and other South Sea Company employees had to evacuate from Spanish territories. After spending most of the war in Europe, Burnet returned to the Americas in 1722 with a new assignment as the South Sea Company's factory surgeon in Cartagena. By 1729 Burnet had grown frustrated by his limited prospects within the company and seized an opportunity to realize his aspirations through espionage. His reward for providing Spain with evidence that the South Sea Company had knowingly and consistently violated the *asiento* contract was a pension and a new position in Madrid. Becoming an agent for the Spanish Crown offered professional and personal rewards Burnet had long unsuccessfully sought in the South Sea Company's employ.[19]

One of the distinctive provisions of the British *asiento* was that it allowed a handful of British subjects employed by the South Sea Company to reside long term in Spanish territories. Although other foreigners resided in Spanish American territories in violation of imperial decrees, the British *asiento* was unusual in providing a legal basis for Britons to live in Spanish colonies for long periods of time. The 1713 *asiento* contract stipulated that up to six British subjects could live at each of the South Sea Company's trading factories. These factors were to "be regarded and treated as if they were Subjects of the Crown of Spain." Burnet and other South Sea Company resident factors were in theory free to trade and travel—and therefore observe and collect—however they wished. Despite the *asiento*'s promises to regard South Sea Company factors as Spaniards, they remained subjects of the British Crown whose allegiances and religion were always suspect. Resident factors' freedom and livelihood remained precarious. They lived under constant threat of Spanish reprisals if the fragile peace between the two empires fractured, as it did twice during Burnet's career. Even in peacetime, South Sea Company factors had to be careful not to run afoul of the Inquisition. The Inquisition's inspection of every book sent by Burnet's correspondents reflected the liminal space the slaving surgeon occupied in Spanish America.[20]

Over the course of Burnet's career in the *asiento* slave trade, he amassed a wide-ranging collection of specimens. The surgeon gave more than 100 specimens to three British naturalists: Sir Hans Sloane, James Petiver, and Dr. James Douglas.[21] The specimens included medicaments, dyes, culinary plants, shells, astronomical observations, and man-made curiosities. About a quarter of them were plants or minerals reported to have medicinal virtues, as one might expect a medical professional to collect. For example, Burnet gathered specimens of *terra macomachi*, a cure for ringworms, from Cartagena; *raiz rouge*, used to stop fluxes, from Buenos Aires; and counter poisons from Jamaica. His professional training meant he also understood the commercial value of imported drugs.[22] Burnet certainly did not confine himself to medicaments. His collection contained over forty animals, including butterflies, a wingless cockroach, three-toed sloths, a marine caterpillar, and a variety of fish. Burnet acquired at least four samples of minerals, including a large amethyst and what he believed was a type of gold. Such specimens manifested British interest in the mineral riches of Spanish America. Similarly, the four specimens of plants renowned as dyes reflected British interest in dyes indigenous to a region already famed for cochineal. And unlike the mineral wealth of Spanish territories, dyes and other types of "green gold" might be transported out of Spanish America and grown in British colonies.[23]

Medicaments and dyes were a frequent focus of efforts to discover "green gold" in Spanish America. Shortly after arriving in Cartagena, Burnet sent Sloane samples of four medicaments popular among local residents. "I should be glad to know if any of these things be Esteemed in England," he wrote, "& whither a quantity of the Earths or Balsam would sell." The South Sea Company factory surgeon frequently complained about the inadequacy of his salary and his limited opportunities to increase his income. He had assumed that he would be able to establish a profitable private medical practice in Cartagena to supplement his South Sea Company salary. Instead, Burnet declared that he had been "deceived in the Place." Cartagena was a city of strategic and commercial importance within the New Kingdom of Granada, yet it was a relatively small city and had not fully recovered from a devasting French raid in 1697. In frustration, Burnet described Cartagena as "the ruins of a good city," which "all the Substantial Merchants haveing forsaken." He claimed that there were few in the city capable of paying the fees for his medical services and that even they were so accustomed to employing Black healers that they paid Burnet so little it was "a disgrace for a Graduate Physician to accept." The minerals and medicaments Burnet sent to Sloane might have solved his financial troubles if, like other medicines imported from Spanish territories, they had commanded high prices in Britain.[24]

Burnet's correspondence does not provide many hints about how he acquired the minerals, medicaments, and other specimens that he gave to metropolitan naturalists. He typically only indicated what he had sent, not how he had acquired the objects. There is no indication that he actively collected specimens in the field himself, although that is possible in some cases. However, in other cases it is clear that he relied on other individuals to collect on his behalf. For example, he told Petiver that he had "sent into the Mountains for ferns." Given the typical collecting practices of Europeans throughout the Atlantic World, it is likely that the person who searched for ferns was of either African or Indigenous descent. Free and enslaved individuals of African and Indigenous descent regularly worked as collectors, guides, hunters, and informants in the early modern Atlantic. Moreover, individuals of European descent were in the small minority in both Portobelo and Cartagena. One historian of Cartagena estimated that 90 percent of the city's population in the early eighteenth century were Black, Mestizo, or Pardo. Burnet likely acquired natural knowledge as well as physical specimens from the local Black, Mestizo, and Pardo populations. The surgeon typically included with the specimen a description of how local medical practitioners used the medicaments he sent. Black and Indigenous healers who dominated Cartagena's medical marketplace are Burnet's most likely sources for local natural and medical knowledge.[25]

The ease with which Burnet garnered the "personal love of the people & especially of . . . Royall officers" represented another key collecting strategy for the slaving surgeon. More than most who worked for the South Sea Company, the surgeon had a talent for making friends among the local Spanish elite. This gift for building cross-imperial alliances was even evident to the Court of Directors in London, who praised Burnet's ability for earning "the Esteem of the Spaniards" and thereby providing the company a "Signal Service." Burnet's Spanish colonial friends helped him to gather new specimens and observations of the natural world. The "Padre Guardian of the Franciscans" in Santa Fe de Botogá, for example, began to correspond with Burnet in 1722. The Franciscan, Burnet boasted, was "somewhat of a Virtuoso who has promised me every thing curious which that Country affords." Burnet also made the acquaintance of an Italian chemist residing in Cartagena. The chemist asked for his assistance acquiring the "Moss of a Mans Scull" in order to complete the "Lapis Butleri of Wonderfull Virtues." Although the surgeon had never heard of the wonderful Lapis Butleri, he asked Sloane to send him the necessary ingredients. Just as Sloane and Petiver developed wide-ranging networks of collectors through their circles of acquaintance and correspondence, Burnet built his own latticework of connections to facilitate the collection of natural curiosities and natural knowledge within Spanish America.[26]

The Spaniards whose esteem Burnet earned included Col. Juan de Herrera, chief engineer of Cartagena and a royal mathematician. In the course of his work for the Spanish Crown, Herrera had traveled extensively in Spanish America, including to Chile, Lima, Panama, and Cartagena. The engineer brought his seventeen-and-a-half-foot telescope with him on his travels in order to observe the skies wherever he went. Shortly after Burnet's arrival in Cartagena in 1722, the two scientifically minded men became friends. Herrera agreed to give Burnet a complete copy of his astronomical observations to forward to the Royal Society of London. Burnet believed that this collection, "all the Astronomical observations in this America," would "be very acceptable to the Royall Society."[27] Yet the South Sea Company surgeon never learned the fate of the observations he sent, nor did he receive the gift that he had requested in return. Sloane gave the astronomical observations sent by Burnet to Edmond Halley, a member of the Royal

Society and the British astronomer royal. Halley was able to compare Herrera's observations of Jupiter's moon Io obscuring the planet with similar observations conducted in England. Together these two sets of data allowed Halley to calculate the longitude of Cartagena. As an important Spanish American port city, accurate geographical knowledge could prove useful to British commercial and military planning. Years later, however, Burnet was still inquiring whether Halley and the Royal Society had found the astronomical observations "acceptable or of any use." He apparently never learned that Halley used the Spanish engineer's observations to calculate the longitude of Cartagena and published the results in the *Philosophical Transactions*. Burnet's role in facilitating the transmission of Herrera's observations was also omitted from the published record. Although Halley acknowledged Sloane for giving him the "Packet of Observations" and Herrera as the individual who had made them, Halley made no mention of Burnet's role as a go-between.[28]

The ease with which Burnet moved between empires and the access to Spanish America that the surgeon enjoyed made him a collector of particular value to metropolitan naturalists. In return, Burnet hoped they would intercede on his behalf in London. The slaving surgeon asked Sloane to oversee the purchase of a new chest of medicines and to send him whatever "new books & Pamphletts &c come out."[29] More importantly, well-connected friends in Britain such as Sloane could plead Burnet's case with the South Sea Company's Court of Directors for promotion or leniency. The slaving surgeon relied on Sloane to act as his patron, asking that he "keep me in the good graces of Sr. John Eyles and Mr. Rudge," the company's leaders. Burnet repeatedly asked to be promoted to the better-paid position of factor. Even with Sloane's lobbying on his behalf, Burnet was told each time that company policy forbade a factory surgeon to become a factor. In 1722, shortly after arriving in Cartagena, he tried a different tack. Burnet begged that Sloane "would use your Interests with the Court of Directors for the enlarging my Salary or my advancement in their service, for it is thoroughing away my time to serve for my present salary." Burnet argued that the job of factory surgeon, if faithfully performed, was much more work than that of factor and that it was in the company's interest to compensate him accordingly. "The diligent discharge of a Physicians duty may save the life of seven or eight slaves in each Cargo which otherways might die & that being saved or lost farr exceeds his Sallary." Burnet attempted to obtain a raise by appealing to the directors' understanding of enslaved Africans as valuable commodities. Yet this was no more successful than his other requests.[30]

Despite Sloane's efforts, Burnet never received the increased salary or promotion to factor that he so desired. In one of his last letters from the Americas, the surgeon confided to Sloane that he was "on the meridian" of his age and could not "afford much more time in the West Indies." He therefore beseeched Sloane to "make use" of his "Interest with the Hon'ble Court of Directors ... to ... give me such a station as I may be able to save a Competency to begin the World anew in Britain & end my days in Europe."[31] Burnet's ultimate decision to become an agent for the Spanish can be understood in light of his disappointed ambitions. In 1728 Burnet was appointed to the British delegation at the Congress of Soissons, which was called to negotiate the end of the Anglo-Spanish War. Burnet used the summit as an opportunity to provide the Spanish with crucial testimony supporting their allegations that the South Sea Company had consistently violated the terms of the *asiento* through contraband trading and bribery. Burnet received a pension and a position as a *médico de cámara* in exchange for his testimony. The surgeon's new allegiance to Spain did not preclude his participation in the networks of British science. Burnet continued to correspond with Sloane during the 1730s, sending natural curiosities and reports on the latest scientific activities from his new home in Madrid. In the porous and interconnected empires of the Atlantic World, Burnet found that a fluidity of identity and allegiance offered the best path to achieving the competency he so desired.[32]

ASIENTO COLLECTIONS

On April 16, 1718, Burnet requested that the South Sea Company's agent in Jamaica, a Mr. Haslewood, assist him with an unusual package he was sending from Portobelo: a "Perico ligero" or three-toed sloth. Burnet offered no suggestions for how Haslewood might care for the animal or even what it might eat. Rather, Burnet simply requested that the sloth be sent with Capt. Edward Noll Coward of the slave ship *John Gally* on the return voyage to England.[33] Unlike most of the American natural curiosities sent to Europe, the sloth was still alive. Transporting living specimens across the Atlantic was a notoriously difficult business. Few naturalists in the eighteenth century attempted it, especially with animals, because the success rate was so low and the potential problems so manifold.[34] Yet Burnet trusted that the South Sea Company's network of slaving vessels and agents could facilitate such a feat.

Captain Coward of the *John Gally* had spent weeks at the Portobelo slaving factory, following a particularly deadly middle passage from Whydah and a stop in Jamaica. Coward and his crew would have assisted Burnet and other members of the Portobelo factory in disembarking the enslaved Africans on board the *John Gally* and preparing them to be sold to Spanish colonists. The weeks Coward spent in Portobelo would have been busy ones for Burnet. Not only was 1718 an unusually busy year at the Portobelo factory in general, but the rate of illness among arriving captive Africans was almost twice that of normal levels.[35] Perhaps this explains why Burnet did not give Coward the three-toed sloth while the captain was in Portobelo. Instead, a few weeks later Burnet entrusted the animal to a ship captain engaged in the transshipment slave trade, Captain Bags. Bags commanded one of the South Sea Company's sloops that ferried enslaved Africans and supplies between the company's entrepôt in Jamaica and its Spanish American factories. Bags accepted responsibility for the sloth by writing on the back of Burnet's letter to Haslewood, "Received of John Burnet a live Animal called Perico Ligero which I promise to deliver to Mr Haslewood."[36]

Burnet's plan for transporting the three-toed sloth involved the cooperation of South Sea Company employees on both sides of the Atlantic. Burnet expected that Captain Coward would personally deliver the sloth to Thomas Knapp, a clerk at the South Sea Company's headquarters in London. Burnet expected Knapp, in turn, to deliver the animal to British naturalist James Petiver at his apothecary shop on Aldersgate Street.[37] The sloth, however, never made it to London. Most likely it never left Captain Bags's possession alive. When Captain Coward was later questioned about the missing animal, Coward declared that he never received it and that neither Haslewood nor Bags ever mentioned anything about it to him. But he did recall that Captain Bags "had a Strange Creature which he made a public Shew of at Jamaica" and that died before he left the island.[38]

Although Burnet's plan for transporting the Portobellan three-toed sloth to London failed, it reflected the surgeon's experience that the routes of the *asiento* slave trade could become those of natural history. Burnet believed that he could orchestrate the safe delivery of the "Strange Creature" through the routes of the *asiento* trade because he had done it before. Throughout his years in the employ of the South Sea Company, Burnet sent letters and scientific specimens to Britain on board the same South Sea Company ships that transported captive Africans and with the crucial assistance of the company's agents and factory personnel. Burnet's collecting practices relied upon not only the access to Spanish America provided by the *asiento* slave trade but also its infrastructure. The rhythms, geographies, and personnel of the British slave trade to Spanish America shaped the collecting practices of South Sea Company employees as well as the natural knowledge that resulted from their collections. The *asiento* trade influenced what was collected, where, how, and by whom.

Naturalists in Britain were quick to realize that the profits of a sanctioned slave trade to Spanish America might include natural observations, specimens, and natural knowledges. Petiver, for example, celebrated the potential scientific windfall promised by the British *asiento* before the agreement was even finalized. Reflecting in 1712 on the many difficulties he encountered in his never-ending quest for new specimens, the apothecary declared that "nothing but the glimmerings of Peace & a South Sea Trade gives me hopes of a faint recovery."[39]

Petiver's excitement about Britain's new trade with Spanish America inspired his *Hortus Peruvianus Medicinalis: Or, The South-Sea Herbal* (1715). *The South-Sea Herbal* contained short descriptions and images of sixty-six Spanish American plants renowned for their efficacy as medicines and dyes. As the naturalist trumpeted in the text's subtitle, the sixty-six plants were "*much desired and very necessary to be known by all such as now* Traffick *to the* South-Seas." *The South-Sea Herbal* was designed to facilitate both collecting and commerce. Most of the plants included in the text were already imported into Britain or had the potential to become valuable new commodities. Local medical knowledge would facilitate the commerce in enslaved Africans by keeping both captives and captors alive. But Petiver equally intended his *South-Sea Herbal* to serve as a guide for what to collect by means of the *asiento* trade and a model for his correspondents to imitate. The apothecary told Dover, for example, that the "Dasie *Ragwort* or *Nillque*" depicted in the second tablet of *The South-Sea Herbal* could be found on the "sea shores" near Buenos Aires. He requested that the factor "make diligent enquiry after" the seed of the Peruvian cress violet. Published in the spring of 1715, just as the South Sea Company's factories were being established, *The South-Sea Herbal* reflected British optimism about the scientific and commercial possibilities created by a legal slave trade to Spanish America.[40]

Petiver was not alone in thinking that the *asiento* slave trade to Spanish America would offer unparalleled access to the region's flora and fauna. An anonymous reviewer in the Royal Society's *Philosophical Transactions* noted in 1712 that it was "'tis now hop'd, the *South Sea Trade* may easily discover and bring" over the flower and fruit of *Cinchona*. The reviewer suggested that the access afforded by the *asiento* slave trade would enable Britons to surreptitiously introduce *Cinchona* into British territories and thereby break Spain's monopoly on the valuable antifebrile. A decade later, the English botanist William Sherard confidently declared that he could acquire specimens of any plant he desired that grew in regions of Spanish America where the South Sea Company traded.[41] Metropolitan naturalists like Sherard understood that the award of the *asiento* to the South Sea Company promised not just the riches of a monopolistic slaving contract but also intellectual treasures that the mercantilist policies of the Spanish Crown normally put out of reach.

These hopes were not misplaced. South Sea Company surgeons, slaving agents, and ship captains gathered samples of Spanish American *naturalia*, investigated the means of their cultivation, and collected rare specimens for British naturalists. Slave ship captain George Jesson, for example, gathered enough butterflies off the coast of Buenos Aires to fill multiple collection books and gathered samples of the "ordinary Paraguay tea" that eventually joined Sloane's museum.[42] In 1715 David Patton pledged to collect specimens for Petiver in Veracruz, where he had recently been appointed the factory's surgeon. Similarly, William Toller wrote Petiver from Buenos Aires after his appointment to a new position as the factory's surgeon. Toller offered to investigate Buenos Aires's flora and fauna on Petiver's behalf and to send the apothecary drawings of all that he encountered. Like Patton and Toller, most of the South Sea Company employees who collected specimens for British naturalists were employed as surgeons either on the company's slave ships or as factory surgeons in Spanish American port cities. Surgeons shared a professional interest in and knowledge about the minerals, animals, and especially plants that played important roles within early modern medicine. They were among the very few Britons to both have access to Spanish American territories and the necessary background to exploit fully this access.[43]

Thomas Dover shared the medical training common among South Sea Company collectors, although his appointment as president of the Buenos Aires factory was mercantile in focus. Dover sent Petiver a botanical collection that he explained had been gathered by his "servants." It is unclear whether these servants were enslaved laborers, local Spanish colonists employed by the factory, or the English servants who were among the unauthorized passengers on the *Warwick*. However, we know that Black and Indigenous individuals throughout the Atlantic World frequently located, collected, and preserved specimens intended for metropolitan collectors. Their role is often obscured in contemporary sources although occasionally it is acknowledged in a backhanded manner in the context of distancing the author from any potential problems with the collection. Dover, for example, mentioned the role of his servants in order to reassure Petiver that "if there should be herein any thing unsatisfactory," it was not due to Dover's neglect. Rather, Dover preemptively blamed any problems with the collection on "the

Ignorance ... [of] those ... whom I ordered to collect, such Herbs, Roots, Flowers, Gums, & other Vegetable productions of these Remote parts of the world." Despite Dover's assertions that his servants were ignorant, it was their natural knowledge and know-how upon which he depended in order to fulfill his pledge to obtain specimens.[44]

The vegetable productions assembled by Dover's servants reflected the geographic reach possible for a South Sea Company resident factor interested in acquiring specimens. The collection did not solely consist of specimens from Buenos Aires. It also included botanical and medicinal specimens from Paraguay and a gum from Mendoza on the eastern side of the Andes.[45] The importance of port cities like Buenos Aires within Spain's internal commercial networks increased the opportunities for South Sea Company factors to obtain specimens indigenous to other places within Spanish America. The infrastructure and geographies of the slave trade within Spanish America created additional opportunities for company employees to acquire specimens. Buenos Aires factory surgeon Francis Hall recruited collectors from among British and Spanish surgeons who accompanied caravans of enslaved Africans sent from Buenos Aires into the interior during the 1720s. He provided these slaving surgeons with quires of paper for pressing plant specimens and directions for how to do so. The geographic boundaries of his own collecting efforts expanded in 1726 due to an epidemic. In that year several South Sea Company slave ships arrived in Buenos Aires with severe outbreaks of fluxes, fevers, and smallpox. The Spanish colonial government responded by ordering that enslaved Africans arriving on South Sea Company vessels had to quarantine for two months on the north side of the Río de la Plata. Hall's role as factory surgeon would have required him to spend significant time at the quarantine site. He used this as an opportunity to collect plants indigenous to the northern bank of the river. The slaving surgeon later sent herbaria specimens prepared from these plants to the English botanist William Sherard.[46]

The nature of collectors' employment within the slave trade also shaped where and when they could acquire specimens. Resident factors and factory surgeons like Hall, Dover, and Burnet typically had greater freedom of movement within Spanish territories, a more extensive social network, and simply more time to collect. The fourteen years Burnet resided in Spanish America was a significant aspect in his ability to acquire a large and diverse group of specimens on behalf of his metropolitan correspondents. In contrast, the rhythms of Capt. George Jesson's work as a slaving captain were more varied and the routes he traveled more diverse. Jesson made multiple trips to Spanish America, commanding both slave ships and South Sea Company packet ships. The itinerant nature of his travels as a ship captain would have allowed him to collect in multiple places. Yet his time in Spanish America was relatively brief and his collecting largely confined to the coast.

South Sea Company mariners like Jesson connected correspondents separated by an ocean. South Sea Company employees based in Spanish America such as Burnet and Hall entrusted letters and specimens to ship surgeons and captains working for the company. A few years after Burnet's failed effort to send a living three-toed sloth from Portobelo, he repeated the attempt with a female sloth and her baby. When the animals died before Burnet could arrange their transport, the surgeon decided to send "the old ones skin stuffed & the young one in Spirits." Burnet directed the package containing the two specimens to the attention of Daniel Westcomb, the South Sea Company's secretary in London. Burnet trusted that the company agents, ship captains, and sailors who handled the package on its long journey from South America to Britain would take particular care with specimens addressed to the influential company official. This time Burnet's faith in the infrastructure of the *asiento* trade paid off. With Westcomb acting as an intermediary, the two preserved sloths successfully reached Sloane, who added them to his museum. Collectors such as Burnet relied upon the *asiento*'s commercial infrastructure to facilitate the transportation of their seeds, specimens, and observations back to Britain.[47]

Like most of the specimens and letters Burnet sent to Britain, the sloths' travels included a stop in Jamaica. Burnet instructed correspondents based in Britain to address his mail and packages to the South Sea Company's factors in Jamaica, trusting that they would ensure the items reached him in Spanish America.[48] Within the commercial networks of the *asiento* trade, Jamaica played a uniquely central role. Three-quarters of the 60,000 enslaved Africans the South Sea Company sold in Spanish America were transshipped from the British Caribbean rather than coming directly from Africa. Most of these enslaved Africans passed through the company's entrepôt in Jamaica. Jamaica's centrality to the South Sea Company's operations in the New World was also reflected in the company's internal hierarchies. The company's top-ranking officials in the New World were those stationed in Jamaica. Their senior position reflected the vital importance of Jamaica to the South Sea Company's American operations.[49]

Jamaica played a similarly pivotal role within South Sea Company employees' pursuit of natural history. South Sea Company collectors relied upon the company's agents in Jamaica to arrange transportation for their specimens and to forward the letters and packages sent in return by European naturalists. Slave ship surgeon William Houstoun entrusted his latest collections to the safekeeping of the South Sea Company's Jamaican agents in between his slaving voyages. Houstoun often divided his smuggled plants and seeds between acquaintances living in different parts of the island, hoping that the plants would thrive in at least one of Jamaica's microclimates. Like most servants of the South Sea Company, Burnet spent months in Jamaica over the course of his career. Burnet's collections indicate that he was not idle during this time. At least ten of Burnet's specimens came from the Caribbean, and most of these from Jamaica. Jamaica was even more important to the collecting activities of Houstoun. Over 40 percent of the species described by Houstoun in his unpublished botanical manuscript were plants he observed in Jamaica. The centrality of Jamaica within the geography of the *asiento* slave trade is reflected in the natural historical collections of South Sea Company employees.[50]

The collecting efforts of South Sea Company employees benefited at various times from assistance lent by company officials. Such assistance by agents in Jamaica and elsewhere was not just professional courtesy. Instead, it also reflected the company's own interest in natural history. Spanish American medicaments and dyes commanded high prices in British markets. Therefore, the South Sea Company had good reason to share an interest in natural commodities indigenous to Spanish America. Further, since Spanish buyers could pay for enslaved Africans in cochineal, cinchona, indigo, or other natural commodities, the company's profits might depend on its employees' command of natural knowledge. South Sea Company directors worried that their factors might unknowingly accept inferior or even counterfeit natural commodities as payment. The directors frequently berated factors who misjudged the quality of dyes and drugs they exported to Britain. In 1717 concern over such issues led them to send John Hoskins "to assist our Factory at Vera Cruz in viewing & Examining Cochineal Indico, and other Dying War[e]s & Drugs." Hoskins, the directors explained, was "a Person of Skill and Judgment" regarding dyes and drugs. The company's directors hoped that under his tutelage the factors at Veracruz could learn to distinguish good-quality dyes and drugs from impostors. Hoskins brought with him samples of Spanish American commodities along with strict instructions that any drugs or dyes purchased by the factory's agents needed to be of at least equal quality.[51]

As the Court of Director's instructions to Hoskins suggested, they were primarily interested in natural knowledge relating to plants with commercial value such as medical substances and dyes. Petiver reflected these priorities when he advised the surgeon William Toller in 1716 that "nothing can better or sooner recommend you to the South Sea Company's Favour or service than Communications of this Kind & especially of such Plants, Roots, Grasses, Minerals &c as relate to dying or any Medicinal use." The naturalist informed Toller that another correspondent who had collected dyes and drugs in Spanish America was rewarded by the company for his efforts. According to Petiver, investigating Spanish American dyes and drugs could be a path to preferment and promotion within the South Sea Company. The value of natural historical investigations for the South Sea Company's Court of Directors lay in the chance that they might identify new natural commodities and thereby improve the company's bottom line. The Court of Directors understood that collecting along the routes of the *asiento* could promote British commercial and colonial ambitions.[52]

The seeds, sloths, plants, and other specimens gathered by Britons in Spanish America bear traces of the *asiento* slave trade that made their collection possible. Dyes and drugs feature prominently among such collections, revealing the priorities of both the South Sea Company and British naturalists. The provenance of such specimens, the itineraries they traveled, and their transportation on South Sea Company slaving vessels reflected naturalists' reliance on the geography and infrastructure of the slave trade. The routes and rhythms of the *asiento* trade created opportunities for company employees, like the ones Burnet, Hall, and Jesson exploited, to acquire specimens on behalf of British naturalists. The *asiento* trade not only provided British naturalists with access to Spanish America but also shaped the collections and natural knowledge that resulted from that access.

PROFITING NATURAL HISTORY

The access to Spanish America afforded to South Sea Company employees explains how the oldest known Argentinian plant specimens can be found today at Oxford University. A Mr. Mylam and his successor, Francis Hall, served as factory surgeons at the South Sea Company's trading post in Buenos Aires during the 1720s. Mylam and Hall collected seeds, plants, and other *naturalia* for English botanist William Sherard. Hall, for example, gathered at least thirty-seven different species in the vicinity of Buenos Aires. He also sent Sherard the "Skin of an Amphibious Bird which is called a Paera Nenia it walks upright when alive," a "Tiggers skin," and "a small piece of a silver mine stone." Sherard's herbarium included both specimens prepared by Mylam and Hall in South America and others prepared in England from plants grown from seeds sent by the surgeons.[53]

Sherard bequeathed to Oxford University these specimens along with the rest of his herbarium, library, manuscripts, and £3,000 to endow a chair of botany. The first Sherardian Professor of Botany, Johann Jacob Dillenius, studied South American plants using specimens in Sherard's herbarium. Dillenius characterized Argentinian plants in his *Hortus Elthamensis* (1732) based on the specimens Mylam had sent to Sherard the decade before. According to modern botanist Arturo Burkart, all of the Argentinian plants Mylam sent were new to European science. Mylam's and Hall's herbarium specimens continue to be available for naturalists to study. Today the Oxford University Herbaria includes over fifty specimens that were gathered by the slaving surgeons in Spanish America.[54]

Argentinian plants collected by means of the slave trade and found today in Oxford are among the modern legacies of *asiento* science. The collections acquired by individuals like Mylam and Hall demonstrate the possibilities open to the small group of South Sea Company employees who used their access to Spanish America to survey the region's natural resources, collect natural curiosities, and bioprospect. The observations they made and the objects they acquired shaped the production of natural knowledge in the eighteenth century. Spanish American natural curiosities gathered along the routes of the *asiento* became part of the Royal Society's repository and joined museums belonging to its members and other metropolitan naturalists. Specimens acquired by means of the *asiento* slave trade reached a broader audience through their depiction in published natural histories and the Royal Society's *Philosophical Transactions*. For example, specimens collected by Burnet provided the basis of the essays on *Ipecacuanha* and armadillos that James Douglas presented to the Royal Society.[55] The collecting practices of South Sea Company employees introduced hundreds of rare specimens—including of new species—into British natural historical collections.

The gardens, museums, and publications that were the beneficiaries of William Houstoun's collecting efforts illustrate how British natural history profited from the *asiento* slave trade. Houstoun worked as a slave ship surgeon on board the South Sea Company's sloop *Assiento* between 1728 and 1732.[56] Unlike most ship surgeons in the slave trade, Houstoun never visited Africa. Instead, he worked in the transshipment trade, transporting captive Africans from Jamaica to South Sea Company factories in Spanish America. The transshipment trade brought Houstoun most frequently to Veracruz and Campeche in New Spain as well as occasionally to Havana, Cuba. During the *Assiento*'s voyages to these port cities, Houstoun was responsible for maintaining the health (or at least the appearance of health) of the captives on board in order to facilitate their subsequent sale to Spanish American colonists. Houstoun and the rest of the *Assiento*'s crew returned to Jamaica after each voyage to await their newest cargo of captive Africans.

Houstoun took advantage of his time in New Spain and Cuba to surreptitiously study local plants and, when possible, collect seeds and prepare herbarium specimens. He similarly studied and collected plants in Jamaica during the weeks or months he spent awaiting his next slaving voyage. Houstoun's efforts resulted in an extensive and largely unpublished four-part botanical manuscript. The largest section of the manuscript, "Catalogus Plantarum in America Observatarum," described 671 different species of plants. The slave ship surgeon recorded plants' size, shape, appearance of their leaves and flowers, where and when they could be found, and during which month they produced seeds. Houstoun often also drew a sketch of the plant to accompany his written description, such as that shown in figure 3.3 of the Kodda-pail plant (*Pistia stratiotes*), which he observed in Veracruz. Houstoun declared that he "met with a great many Plants" in New Spain that he believed represented new genera, not just new species. He "made bold to characterize some of them," often naming them in honor of European botanists. The manuscript suggests that Houstoun was particularly interested in plants found in Veracruz and Jamaica. Veracruzan and Jamaican plants account for nearly 90 percent of the specimens described in Houstoun's manuscript.[57]

The seeds, cuttings, and roots of plants Houstoun collected during the course of his employment in the *asiento* slave trade were destined for metropolitan collections and gardens. In 1730, for example, the slave ship surgeon sent Sloane "a Collection of Plants and other natural Curiosities from La Vera Cruz."[58] This collection would have been gathered during one of the *Assiento*'s trips to New Spain to transport captive Africans. Sloane's museum included many items sent by Houstoun from Veracruz, although it is impossible to determine which were gathered during this particular trip to the Spanish American port city in 1730. Among the Veracruzan specimens Sloane credited Houstoun with contributing to his museum were the fruit of a tulip tree, pods of the *Acacia americana*, and the stalk of a fern.[59] Historian Victoria Pickering's study of Sloane's *Vegetable and Vegetable Substances* catalog indicates that Sloane credited Houstoun with contributing botanical specimens at least eight-eight times. Forty-five of those entries reference collections Houstoun made in Spanish America, while the remainder were gathered in Jamaica.[60]

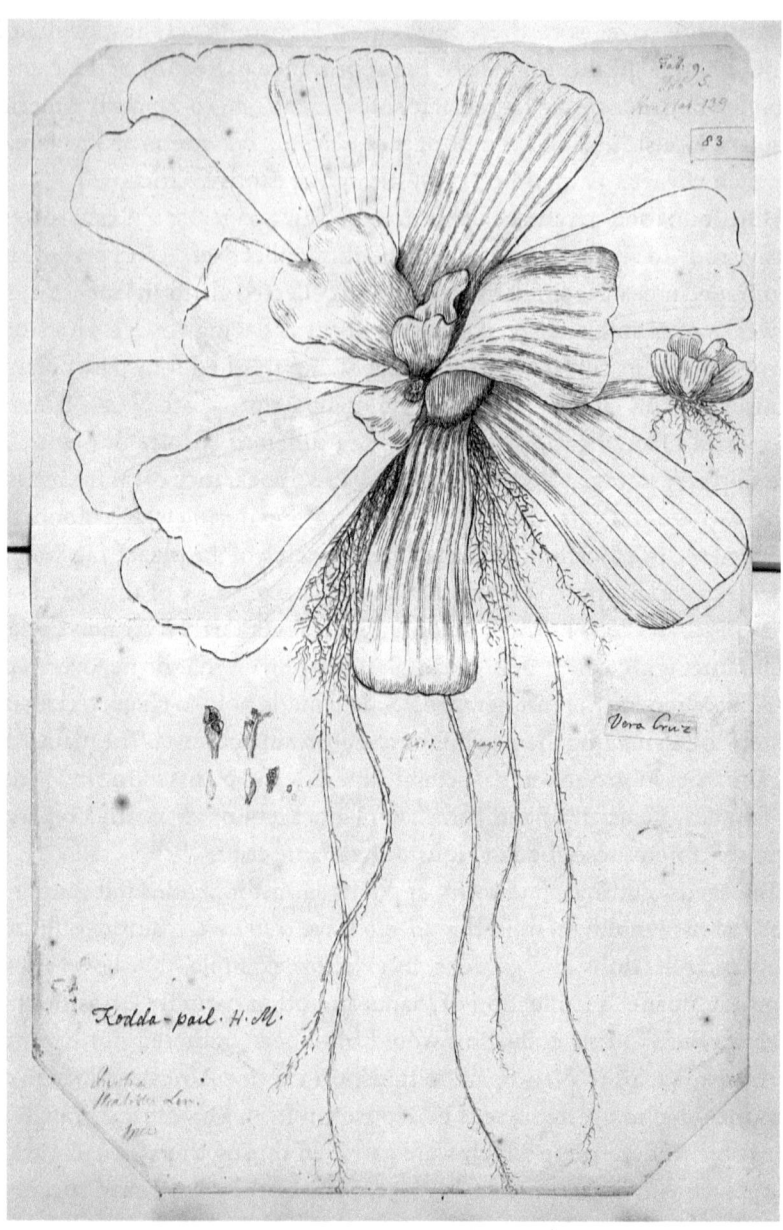

FIGURE 3.3. Kodda-pail plant (*Pistia stratiotes*). Slave ship surgeon William Houstoun described and drew hundreds of plants in Jamaica and New Spain while he was employed in the transshipment slave trade between British and Spanish territories. These botanical drawings include this image of the Kodda-pail plant from Veracruz. William Houstoun, "Drawings of Plants by William Houstoun from Central America, chiefly from Vera Cruz," MSS Banks Coll Hou, Natural History Museum, London. © The Trustees of the Natural History Museum, London.

Another frequent benefactor of Houstoun's collecting efforts was his close friend Philip Miller. Miller was the head gardener at the Chelsea Physic Garden located just outside of London as well as a member of the Royal Society of London and author of the popular *Gardeners Dictionary*. Whenever possible, Houstoun gathered seeds or roots of Spanish American and Jamaican plants for Miller to grow in the Chelsea Physic Garden and sent dried plants with additional information about the specimens. In the eighth edition of *The Gardeners Dictionary*, Miller credited Houstoun with introducing 289 species of plants into the Chelsea Physic Garden, more than 200 of which Houstoun gathered in Spanish America. Of these, all but 25 are likely to date from the three years when Houstoun was employed as a slave ship surgeon. This marks an extraordinarily large number of new plants to be introduced into English botany in a relatively short period, and all through the exploitation of the routes of the slave trade. The large number of plants gathered by Houstoun helps to contextualize botanist Richard Pulteney's comment at the end of the eighteenth century that the diversity of plants found in the Chelsea Physic Garden during Miller's tenure as the head gardener was "particularly" indebted to Houstoun's collecting efforts.[61]

Houstoun's collecting efforts as a slave ship surgeon continued to shape botany after his death, in part because he bequeathed his manuscripts, drawings, and herbarium to Miller. Nearly two decades after Houstoun's death, Miller was still actively using Houstoun's papers in the course of his own botanical studies. Miller's *Gardeners Dictionary* referenced Houstoun's unpublished botanical manuscript more than 180 times. When the gardener presented the Royal Society in 1753 with specimens of previously undescribed American plants, he included drawings made by Houstoun in the Americas twenty years earlier.[62] Miller also allowed other naturalists who visited the Chelsea Physic Garden to consult Houstoun's papers. For example, when Carolus Linnaeus visited the garden in 1736, he had access to both Houstoun's manuscripts and living plants grown from seeds he had sent.[63] Cambridge professor of botany John Martyn described in his *Historia plantarum rariorum* fourteen Spanish American plants raised from seeds or cuttings sent by Houstoun. Martyn's work shows evidence of studying both the living plants in the Chelsea Physic Garden and Houstoun's manuscripts. Martyn noted with approval, for example, that the roots of the *Sinapistrum zeylanicum* grown in Chelsea looked precisely how Houstoun had drawn them in his unpublished botanical manuscript. The Cambridge botanist quoted wholesale from Houstoun's unpublished descriptions of Spanish American plants. Other botanists, including Isaac Rand, director of the Chelsea Physic

Garden, and Adrian van Royen, professor of botany at the University of Leiden, similarly cited Houstoun's manuscripts in their own publications.[64]

Many of Houstoun's papers and herbaria specimens survive in modern museums and libraries. Sir Joseph Banks acquired Philip Miller's herbarium after the gardener's death in 1771. Houstoun's manuscripts, drawings, and herbarium specimens composed a small part of Banks's vast natural historical collections, which joined the British Museum after Banks's own death. These specimens ultimately became part of the collections of the Natural History Museum in South Kensington when the museum was created out of the British Museum as a separate institution in 1881.[65] Other specimens collected by Houstoun joined the herbaria of the surgeon's friends, patrons, and contemporaries. For example, ninety-six Houstoun herbaria specimens can be found scattered throughout the twelve-volume herbarium originally owed by Robert James, 8th Baron Petre. These specimens are held today by the Sutro Library in San Francisco.[66]

Houstoun's specimens are just one example of how objects and knowledges collected through the routes of the slave trade to Spanish America survive in natural history museums, university herbaria collections, botanic gardens, and libraries. British natural history also benefited from specimens that do not survive in modern collections like the butterfly books filled by slaving captain George Jesson. The *asiento* slave trade made possible Toller's natural history of the voyage to Buenos Aires, Houstoun's "Catalogus Plantarum," and Petiver's published descriptions of Spanish American flora and fauna. These natural historical texts and specimens remain a valuable resource for those interested in taxonomy, biodiversity, and any number of related questions. They continued to influence the production of natural knowledge long after the *asiento* ended. Such legacies suggest that we should count Veracruz plants, smuggled jalap roots, preserved sloths, and books of butterflies among the proceeds of the British slave trade to Spanish America.

CONCLUSION

Museums, gardens, and herbaria do not typically come to mind when we think about the profits of the slave trade. Yet British naturalists exploited the commercial networks of the early eighteenth century slave trade to Spanish America in order to acquire specimens that otherwise were difficult or even impossible for them to obtain. South Sea Company employees like John Burnet, George Jesson, and William Toller occupied a liminal and relatively unique space as British subjects allowed to visit or reside in Spanish territories because of their employment in the slave trade. For those interested in exploiting it, the access created by the *asiento* slave trade enabled natural historical collecting at the interstices of empire. The geographies, personnel, exigencies, and infrastructures of the *asiento* slave trade shaped the collections and natural knowledge that resulted. Collecting along the routes of the *asiento* was not incidental to the trade. Instead, it relied on structures and systems specific to the British slave trade to Spanish America. Specimens acquired by South Sea Company employees like Burnet and Jesson challenge us to think expansively about slaving's profits. Not all proceeds of human trafficking can be tallied in a merchant's ledger book. Specimens, seeds, sloths, and astronomical observations may be less easily quantified but represent profits of the British slave trade nonetheless.

The pursuit of profit in part explains the fluidity of the early modern Atlantic World. Boundaries between rival European empires appear fixed on contemporary maps. On the ground, however, borders could be more porous especially when commercial opportunity beckoned. For some individuals like John Burnet, identities and allegiances became fluid when the right opportunity occasioned it. The porousness of the Atlantic World was even more pronounced in the Greater Caribbean, where smuggling and contraband trading flourished. Some estimates suggest that illicit trade surpassed that of legal commerce in value.[67] Plants, seeds, butterflies, and other specimens smuggled on board South Sea Company vessels can thus be understood as part of this interimperial contraband trade. The natural historical profits of the *asiento* were largely illicit.

Although the collecting activities of South Sea Company employees like Burnet and Jesson were illicit, their presence in Spanish America was aboveboard. In the next chapter we see how a South Sea Company slaving surgeon returned to Spanish America under the cover of the *asiento* trade. William Houstoun's extensive experience collecting and observing plants in Spanish America while an employee of the South Sea Company inspired him to attempt to return to the region as a full-time plant collector. He and his successor, Robert Millar, sought to collect rare plants along the routes of the *asiento* slave trade on behalf of a group of British patrons without actually being employed by the South Sea Company.

4.

Botany under the Cover of the Slave Trade

In March of 1732 William Houstoun counted himself lucky. It had been a difficult winter for mariners plying the Gulf of Mexico and the Caribbean Sea, with unusually strong winter storms damaging or destroying many vessels. As a slaving surgeon engaged in the transshipment trade to Spanish America, Houstoun frequently crisscrossed this portion of the Atlantic. In February his vessel had encountered a fierce storm and sank near Veracruz. Houstoun, however, escaped unscathed and had "the good luck to save most of what belonged" to him. Despite Houstoun's good luck, the destruction of his ship meant that he was out of a job. As he explained to the British naturalist and collector Hans Sloane, "the loss of my business oblidges me again to have recourse to you begging that you'd use your interest a second time to put [me] in a way of liveing capable to serve you." Houstoun initially requested that Sloane help him to secure another position as a slave ship surgeon with the South Sea Company. By the time Houstoun made his way back to Britain, he had hit upon a new solution. Rather than return to the slaving company's service and collect natural historical specimens on the side, he would become a full-time plant collector, engaging in bioprospecting under the cover of the slave trade.[1]

The years Houstoun spent as a South Sea Company surgeon had taught him that the British slave trade to Spanish America provided Britons with access to Spanish territories. The Asiento de Negros gave its holder the monopoly on the legal slave trade to Spanish America. Houstoun's employer, the British South Sea Company, held the *asiento* between 1713 and 1739. The slaving surgeon was among the handful of South Sea Company employees to exploit this access in order to illicitly gather seeds, plants, specimens, and observations of rare and desirable plants. After the sinking of the *Assiento*, Houstoun planned to return to the Americas under the pretense of being employed by the South Sea Company. He proposed that the South Sea Company's vessels would provide his transportation to port cities in Spanish territories and that the trade in captive Africans would provide the excuse for his presence. He would be free to devote all his time to searching Spanish America for flora from which drugs, dyes, and other valuable commodities might be obtained. Houstoun, however, died less than a year into his proposed three-year expedition. It fell to his successor, Robert Millar, to bioprospect along the routes of the *asiento* trade.

Bioprospecting had been part of the British slave trade to Spanish America from the beginning. South Sea Company employees clandestinely investigated Spanish American *naturalia*, gathered specimens, and transported them by means of the infrastructure of the slave trade. Yet what Houstoun proposed in 1732 was different. Earlier collecting efforts along the routes of the *asiento* were undertaken by individuals authorized to visit Spanish America by virtue of their involvement in the slave trade. Houstoun proposed to return to Spanish America illicitly with only the pretense of being engaged in the *asiento* slave trade. Further, earlier bioprospecting efforts by South Sea Company employees were often haphazard, while Houstoun's plan represented a systematic attempt to collect in Spanish America by means of the slave trade. His plans constituted a concerted effort to exploit the commercial networks of the British slave trade to Spanish America in order to introduce new plants into British gardens, to promote British natural history, and to encourage British colonialism, particularly in the new colony of Georgia.

Long before Britain held the *asiento*, British merchants and imperial officials understood that the slave trade provided an opening to Spanish America that could be further exploited. The demand for enslaved laborers in Spanish America led Spanish officials to establish the *asiento* trade as a legally sanctioned exception to commercial policies that on paper strictly forbade foreign traders. Further, local officials were often willing to countenance illicit trade that violated imperial policies if it supplied enslaved laborers. As historian

Gregory O'Malley argued, the trade in captive Africans "opened the door" to Spanish America. The South Sea Company exploited this by ordering vessels to take on enough enslaved Africans at its Caribbean entrepôts to qualify the voyage as a slave ship and thus gain admittance to Spanish American ports. O'Malley observed that "the slave trade often served as cover for a broader, illegal trade in British manufactures and other commodities."[2] Smuggling was endemic within the slave trade to Spanish America. Enslaved Africans, manufactured goods, and provisions were frequently smuggled into Spanish America by means of the slave trade while bullion and colonial products were smuggled out.[3] Houstoun's proposal to bioprospect under the pretense of being engaged in the slave trade highlights another form of contraband smuggled out of Spanish America. The proposal drew upon Houstoun's previous experience successfully smuggling plants and natural knowledge by means of the commercial infrastructure of the *asiento* slave trade.

CONTRABAND COLLECTING

During the years that Houstoun worked as a slave ship surgeon for the South Sea Company, he illicitly collected, observed, and studied hundreds of American plants. The seeds, drawings, herbarium specimens, and botanical observations he made enriched the collections of metropolitan naturalists and gardeners. They also provided the basis for his contributions to the Royal Society's scientific meetings and the justification cited in his election to the society's membership.[4] The surgeon was among the most prolific of eighteenth-century Britain's collecting slave traders. Houstoun understood from personal experience that an employee of the South Sea Company could turn the access to Spanish America created by the slave trade into rare collections, new natural knowledge, and unexpected personal opportunities. To do so, he relied on the appropriation of Indigenous, African, and Pardo knowledge, know-how, and labor.

Houstoun served as a ship surgeon on board the South Sea Company vessel *Assiento* from approximately 1728 until early 1732. The *Assiento* traveled throughout the Caribbean Sea and the Gulf of Mexico, ferrying enslaved Africans from the South Sea Company's entrepôt in Jamaica to its trading factories in Spanish America. The African men, women, and children on board the *Assiento* would have already endured the horrors of the middle passage. Upon arrival in Jamaica, they would have found themselves brought up on deck, where South Sea Company's factors subjected them to further dehumanizing treatment as they inspected them for illness and general health.

Spanish buyers had a reputation for only accepting the healthiest and most desirable captives, so those captives deemed to be too ill, maimed, or old for the Spanish market would be sold at discount prices in Jamaica. The rest were transshipped to Spanish American ports on vessels like the *Assiento*, which typically carried smaller cargoes of captive Africans. It fell to Houstoun to ensure that African captives remained healthy during the voyage, or at least would appear so to Spanish buyers.[5]

In December 1730 Houstoun apologized for his collecting efforts by reminding Sloane that "you cannot expect me to do so much as if I lived ashore, since you know the many hinderances that one must necessarily meet with in a sea faring life and especially in a small vessel."[6] Sloane would have understood that the surgeon had it backward. Houstoun's seagoing life was precisely what enabled his bioprospecting and made him such a valuable correspondent. Houstoun's ability to observe, describe, obtain, and transport hundreds of rare American plants was inextricably linked to his employment as a mariner in the *asiento* slave trade. Yet the role played by Houstoun and other mariners in the production of natural knowledge during the early modern period is often absent in modern scholarship. The national—and largely terrestrial—perspective that dominates most historical scholarship tends to position mariners as incidental figures, itinerants moving in and out of the main scene of action.[7] Recentering our focus onto the geography inscribed by eighteenth-century commercial routes rather than the geographies determined by modern political boundaries brings into focus individuals like Houstoun. For a slaving mariner, the Atlantic World was more porous than imperial boundary lines on the map might suggest. For a British slaving mariner interested in natural history, such fluidity provided matchless opportunities to acquire and observe Spanish America's *res naturae*.

As Houstoun complained, the *Assiento* was a small ship, perhaps half the size of a typical British transatlantic slaving vessel. On an average, the *Assiento* carried just under a hundred captives on each of its voyages to New Spain.[8] The smaller size suited the *Assiento*'s purpose in the more local transshipment trade between Jamaica and Spanish territories. Despite the close quarters, Houstoun's belongings would have included seeds, detailed sketches of Spanish American plants, and manuscripts detailing the medicinal and commercial properties of American flora. Houstoun described in his unpublished botanical manuscript nearly 700 American plants, most of which he observed in Veracruz or Jamaica. The surgeon's baggage would have swollen during return voyages to Jamaica to include boxes of living plants smuggled out of Spanish territories. Notwithstanding the occasional setback, Houstoun reported that he "met with a great many Plants" in Spanish America. Countless of these, he thought, were new to European science, but without access to a botanical library, he was never confident they were truly nondescript.[9] Although the slave ship surgeon occasionally botanized in Havana and Campeche, the majority of Houstoun's bioprospecting in Spanish America—and likely the majority of his slaving—occurred in Veracruz.

Veracruz was the principal port connecting New Spain with the Atlantic World and with Spain itself. It was the gateway through which passed the legendary wealth of New Spain. By imperial decree, it was the only legal port of entry for colonial Mexico. European manufactured goods, provisions, free and unfree migrants, and colonial products such as silver, gold, and cochineal all funneled through Veracruz. It was "the most important port of the richest possession within the Spanish colonial empire," yet due in large part to environmental factors, it did not experience urban growth commensurate with its commercial importance. Veracruz had one of the largest Black populations in colonial Mexico, which was a legacy of the extensive use of enslaved labor in domestic service, urban trades, and the region's sugar industry beginning in the sixteenth century.[10]

Slavery and the slave trade played a pivotal role in colonial Veracruz's development, but its importance was on the wane when Houstoun visited in the 1730s. Veracruzanos turned to enslaved Africans as a labor source by the mid-sixteenth century following the catastrophic decline of the local Amerindian population. By 1600, individuals of African descent outnumbered Europeans in the region. Most Black Veracruzanos in this period hailed from the various cultures of West Central Africa. Veracruz in the early eighteenth century continued to have a significant Black population, but growing numbers of Amerindian, European, and mixed-race individuals (Pardo, Mestizo, and Mulatto) meant that Black Veracruzanos were no longer in the majority. By the eighteenth century, most Afro-Veracruzanos would have been born in New Spain. The arrival of new *bozales*, enslaved Africans, was a relatively rare occurrence after 1640, but these new arrivals introduced some ethnic diversity within the region's Black population. Captives transported by the South Sea Company included individuals from the Gold Coast, West Central Africa, the Bight of Benin, and the Bight of Biafra.[11]

The Veracruz visited by Houstoun in the 1730s contained significant Indigenous, Black, and mixed-race populations, who by this period were more likely to be wage laborers than enslaved. The recovery of the Indigenous population, general population growth, and an unreliable supply of new captive Africans contributed to the declining importance of slavery in Veracruz in the half century before Houstoun visited the city. In part, this explains why the South Sea Company sold far fewer captives in the Mexican port than in any other city where it had a factory, despite the immense commercial importance of Veracruz and the region's wealth. More than six times as many captive Africans were sold by the Portobelo and Panama factory than by that in Veracruz.[12] The greater commercial significance of the South Sea Company's commerce in Veracruz lay in the contraband trade it enabled. It also made possible Houstoun's contraband collecting.

The nature of the transshipment slave trade meant that Houstoun undertook multiple voyages to Spanish America each year. In 1730, for example, the slaving surgeon made at least three separate voyages from Jamaica: two to Veracruz and the third to Campeche, on the Yucatán Peninsula. During the weeks that the *Assiento* was in a Spanish American port, Houstoun would have been responsible for the medical care of the captives on board the vessel as well as intimately involved in preparations to sell them to Spanish colonists. During this period Houstoun also botanized in the surrounding countryside within a day's journey in search of new, rare, and potentially valuable plants.[13]

Houstoun does not indicate who accompanied him on these bioprospecting trips, but the collecting practices of contemporaries suggest that he likely relied on local guides, hunters, specimen collectors, and informants of Indigenous or African descent. Traveling naturalists often portrayed themselves as heroically surviving the rigors of the field unaided. In reality, local knowers and laborers made such exploits possible. Individuals of African, Amerindian, and mixed-race descent frequently guided naturalists' travels, collected and carried their specimens, packed their seeds and plants, and informed them of local uses of flora and fauna. The British naturalist Mark Catesby, for example, hired Native Americans as hunters, guides, specimen collectors, and assistants throughout his travels in South Carolina in the 1720s. Catesby declared himself "much indebted" to the "Friendly Indians" whom he "employ'd . . . to carry my Box, in which, besides Paper and Materials for Painting, I put dry'd Specimens of Plants, Seeds, &c." His Indian assistants also guided his journey, hunted for his food, and gathered particularly illusive specimens. Two decades earlier the English naturalist and apothecary James Petiver instructed his young apprentice George Harris to train the enslaved Africans he met on his travels "how to collect things by taking them along with you when you are abroad."[14] Houstoun's near contemporary Charles-Marie de La Condamine relied on the knowledge, labor, and skills of Amerindians and enslaved Africans during his expedition to Quito to measure the curvature of the earth. Yet little evidence for this appears in sources that the French natural philosopher himself produced. As historian Neil Safier concluded regarding the Quito expedition, "printed texts tend to paper over the processes by which they came into being," effacing the intellectual and physical contributions of a range of social actors, especially individuals of Indigenous and African descent.[15]

In May 1730 the *Assiento* arrived in Veracruz with 117 captive Africans transshipped from Jamaica. During the weeks the vessel was in port, Houstoun botanized outside the nearby settlement of Old Vera Cruz (now Antigua), located a few leagues from the harbor. The surgeon reported that he observed a plant growing on the high ground along the Antigua River that he believed was sold in British apothecary shops as the drug *contrayerva*. Like most of the seemingly miraculous medicines imported from abroad, relatively little was known about *contrayerva* in Britain. Common wisdom held that it was "very good against all sorts of malignant Fevers, and pestilential Distempers; resists Poison, and the Bites of venemous Creatures." In spite of its long popularity, British physicians, apothecaries, and naturalists alike were unsure from which plant it derived or even where it grew. Most Britons believed *contrayerva* referred to a particular plant. Houstoun learned, presumably from local informants, that in Spanish America it referenced an entire category of herbal medicines: those good against poisons. Despite the many plants that shared the name, Houstoun believed that the particular plant growing outside of Old Vera Cruz represented one of two plant species sold as *contrayerva* in British apothecary shops.[16]

The transshipment of captive Africans brought Houstoun to Campeche later that same year, in November of 1730. On "high rocky grounds about Campeche," Houstoun encountered the other plant he believed was typically sold in Britain as *contrayerva*. Houstoun carefully noted the botanical characteristics for both plants and sketched them on the spot, just as he had done with scores of other plants. His surviving manuscripts provide no indication about how local knowledge may have informed his choices about where to botanize or whether (as seems likely) a local guide led him to the valuable plants. The potential importance of the medicament led Houstoun to acquire living samples of both plants and to smuggle them out of Spanish America on board the *Assiento* in the hope that he could transplant them into Jamaican gardens. In Vera Cruz this required that he purchase "four plants of it in earth." The surviving historical record does not indicate from whom he made this purchase. Other eighteenth-century bioprospectors such as Nicolas-Joseph Thiéry de Menonville purchased desired specimens from Black and Indigenous proprietors. The specimens Houstoun acquired in Veracruz were destroyed during the voyage by a severe storm. However, the surgeon was able to introduce plants he believed to be *contrayerva* from Campeche into Jamaican gardens.[17]

For Houstoun personally, the introduction of *contrayerva* plants into Jamaica was not as significant as the publication of his account of having done so. In 1731 Houstoun sent Sloane "An Account of the Contrayerva," his botanical description of the two plants collected by means of the transshipment slave trade to New Spain the previous year. Sloane read the paper to the Royal Society of London on Houstoun's behalf. In it the slaving surgeon argued that the drug known in London as *contrayerva* came from the roots of the two plants he had acquired in New Spain. These, he explained, were both species of the *Dorstenia* (see fig. 4.1). Based on "An Account of the Contrayerva," Houstoun was credited within European botany for having discovered the botanical source of *contrayerva*. No credit was given to the local informants and guides whom Houstoun relied upon and from whom he likely appropriated botanical knowledge about the *Dorstenia*. The subsequent publication of "An Account of the Contrayerva" in *Philosophical Transactions* became the definitive account of the medicament, cemented Houstoun's reputation as a botanist, and led to his election to membership in the Royal Society.[18]

Houstoun was able to view the *Dorstenia* in person because the plants grew within a few hours' walk from where the *Assiento* lay at anchor. The other popular medicament Houstoun acquired in Veracruz in 1730, jalap root, grew further inland. Houstoun turned to Indigenous collectors in order to obtain the valuable drug. Jalap root (*Ipomoea purga*) was a staple of eighteenth-century British apothecary shops. The mild purgative's importance in Veracruz predated the arrival of the Spanish in the sixteenth century. The Aztec (or Mexica) had demanded the drug, known as *mechoacán*, as tribute from the local subject peoples. Spanish colonists renamed the drug *purga de Jalapa* after the region where it was commonly found and from where it remained an important export throughout the eighteenth century.[19]

Despite the drug's popularity, British apothecaries and naturalists were uncertain from which plant jalap root was derived. Houstoun initially thought that since "all the Jallap is exported from Vera Cruz," he would be able to determine the medicament's botanical identity during one of the *Assiento*'s trips to sell African captives and deliver supplies to the company's factory in New Spain. But to his disappointment, he discovered he "could learn nothing" in Veracruz concerning the botanical identity of jalap. Undeterred, the surgeon vowed to visit the eponymous region where the plant was grown if

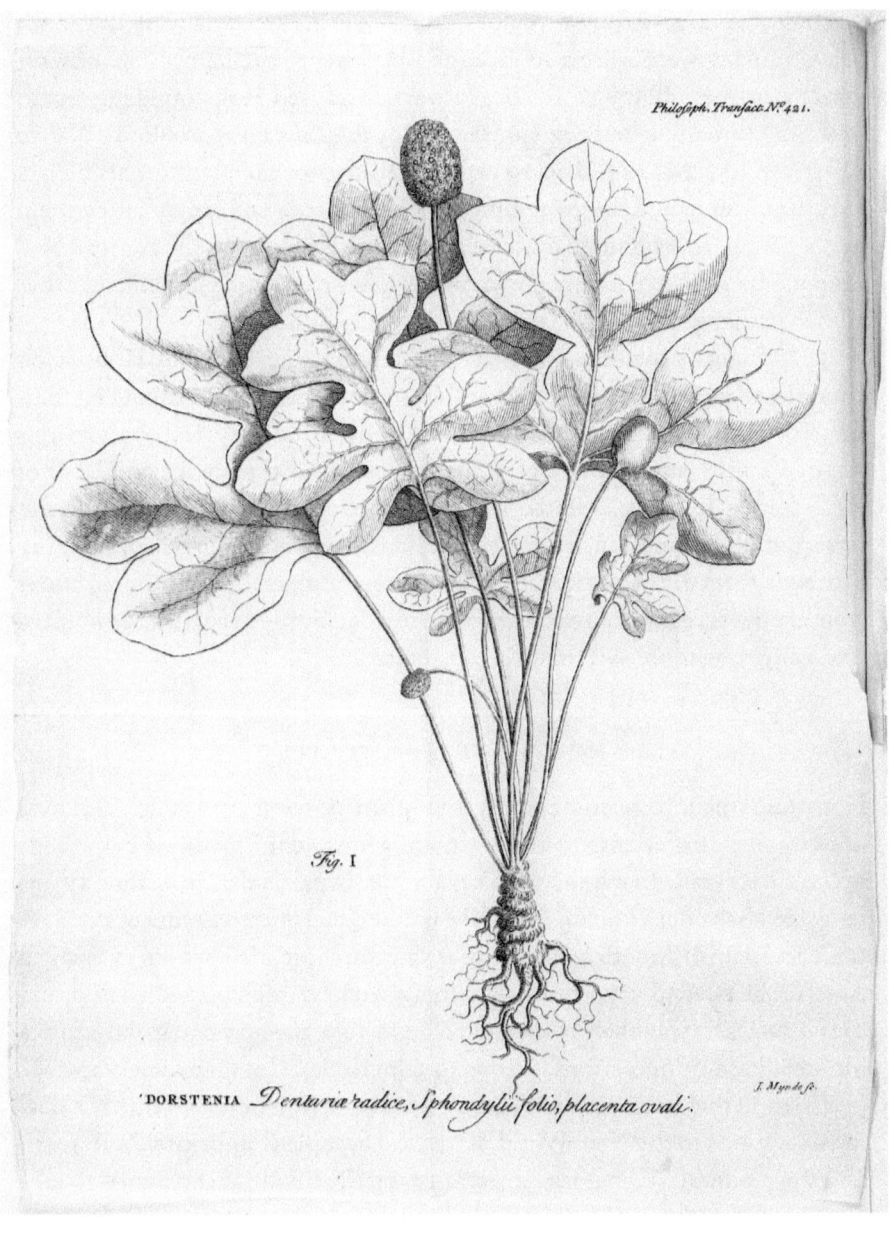

FIGURE 4.1. *Dorstenia*. William Houstoun identified the popular medicine known in Britain as *contrayerva* as the *Dorstenia*, indigenous to New Spain. He observed the plant while a slave ship surgeon in the employ of the South Sea Company in 1730. William Houstoun, "An Account of the Contrayerva, by Mr. William Houstoun, Surgeon in the Service of the Honourable South-Sea Company," *Philosophical Transactions* 37 (October–December 1731): foldout opp. 184, fig. 1, RB 750073, Huntington Library, San Marino, California.

he could "have leave from my Superiors." While his supervisors in the South Sea Company were willing to give their blessing to such a trip, Spanish officials were less obliging. The local governor refused Houstoun's request to make the three-day journey. Consequently, the ship surgeon hired a Native American to travel to Xalapa to gather seedlings of the plant on his behalf. Houstoun smuggled these seedlings out of Veracruz and transplanted them into a garden belonging to a friend in Jamaica. Seeds from the transplanted jalap plants were eventually grown in the Chelsea Physic Garden and other British gardens.[20]

Jalap plants grown in Chelsea were among the profits of Houstoun's contraband collecting while a slave ship surgeon. Houstoun smuggled hundreds of seeds, plants, botanical drawings, and natural historical observations out of Spanish America on board the *Assiento*. The knowledge and labor of local guides, hunters, collectors, and informants, such as the Indian who collected jalap plants, underwrote the slaving surgeon's botanical profits. The proceeds from Houstoun's contraband collecting inspired him to consider whether the access to Spanish America provided by means of the *asiento* slave trade might be more systematically exploited.

HOW TO BE A BIOPIRATE

Houstoun's plan to become a full-time plant poacher in Spanish America drew upon his experience as a slave ship surgeon and the firsthand knowledge of the *asiento* trade that he acquired as a result. In particular, Houstoun's plans reflected an insider's knowledge of the commercial infrastructure of the slave trade to Spanish America. The former ship surgeon knew where in Spanish America he needed to go, how to get there, what he might expect to find, and the obstacles he was likely to encounter. Houstoun proposed to make Jamaica his local base of operations, just as the South Sea Company itself did. He understood that because Jamaica was the central hub of the company's slaving operations in the New World, it was the best place in British America to find ships bound for Spanish American ports. The naturalist planned to take passage on South Sea Company ships traveling to the company's factories in Cartagena, Portobelo, Campeche, and Veracruz. In between each voyage the naturalist would return to Jamaica. There Houstoun would leave any plants he collected with South Sea Company agents and other colonists "capable and willing to take care of them." Eventually the plants he collected along the routes of the *asiento* could be transplanted into the new colony of Georgia and other British territories.[21]

The introduction of Spanish American plants into British territories benefited avid gardeners and botanists. But the larger significance of Houstoun's smuggled flora was rooted in mercantilism. The mercantilist economic philosophy shared by European powers in the eighteenth century taught that natural commodities could play a crucial role in the creation of a nation's wealth. This philosophy stressed the need of "selling more to strangers yearly than we consume of theirs in value," as one seventeenth-century English writer explained. Acquiring domestic sources for imported natural commodities contributed to a positive balance of trade. Colonies could then supply raw materials that the home country would otherwise need to import from its rivals, thus preserving the nation's supply of bullion. Early promoters of British imperialism predicted that American colonies would provide the citrus, olives, sugar, silk, medicines, dyes, and other natural commodities that Britons purchased from other nations. While British American colonies turned out not to be well suited to many of the drugs, dyes, and foodstuffs most desired, the establishment of the new colony of Georgia in 1732 rekindled these discussions. Georgia's promoters argued that the new colony's latitude, climate, and proximity to New Spain would enable it to grow many of the semitropical plants that had not thrived in other British colonies.[22]

Mercantilism encouraged imperial powers to guard the natural knowledge and natural resources of their own dominions against the bioprospecting efforts of their rivals. Efforts to study and to poach rivals' natural commodities were almost as old as European colonialism itself. Many of these clandestine efforts crossed from bioprospecting into biopiracy. Today "biopiracy" is most often used to describe intellectual property disputes between multinational pharmaceutical companies and the Indigenous groups whose traditional knowledge they seek to patent.[23] Historian Londa Schiebinger borrowed the language of biopiracy to describe early modern naturalists who searched for natural commodities while in a foreign territory under false pretenses.[24]

The importance of the *asiento* to Houstoun's plans to become a biopirate went beyond providing convenient conveyance to Spanish America. More fundamentally, the naturalist believed that the routes of the slave trade were the only way to gain access to Spanish dominions. In his proposal, Houstoun stressed that his plan required the assistance of the South Sea Company. He was concerned that the company's Court of Directors in London should support the plan and, more crucially, that they should instruct their factors in the Americas to do so. He singled out the need for the cooperation of Edward Pratter and James Rigby, the South Sea Company's agents in Jamaica. It would be up to Pratter and Rigby to "grant him his passage on board the Company's

Vessels" bound for Spanish America. Similarly, Houstoun asserted that the company's factors in Portobelo needed to devise "some pretext or the other" for sending him across the isthmus to Panama "because unless he goes as the Company's Servant he will not be allowed by the Spaniards to cross the Countrey." Houstoun understood that without the cooperation of South Sea Company agents, his biopiracy expedition would be over before it ever began. The former surgeon's plan reflected his knowledge of the physical and political geographies of the British slave trade to Spanish America.[25]

Houstoun's South Sea Company experience alone was insufficient to pull off the expedition he envisioned. He needed the help of well-connected friends to secure the financial and logistical support his plans required. For this, he turned to the two men who knew better than anyone his skill and tenacity as a collector: Hans Sloane and Philip Miller. Sloane and Miller had been the primary recipients of the hundreds of pressed plants, seeds, and roots that Houstoun collected while a slave ship surgeon employed by the South Sea Company. During the first half of the eighteenth century, Sloane was the most influential patron of science in Britain. As a royal physician, natural historian of Jamaica, and president of both the Royal Society of London and the Royal College of Physicians, Sloane had the connections and resources necessary to make Houstoun's plans come to fruition. Although Miller did not enjoy Sloane's wealth or elite social status, his reputation as the most talented gardener of his generation brought him into contact with many of the most avid patrons of botany in Britain. He was best known as the author of *The Gardeners Dictionary*, the first comprehensive guide to practical gardening. Contemporaries admired Miller's simple, unadorned language, practical advice, and straightforward descriptions of each plant. According to one contemporary, *The Gardeners Dictionary* was the one book every gardener in Britain consulted daily.[26]

Sloane and Miller were likely involved with the plan to bioprospect in Spanish America from its inception. They may have even been responsible for suggesting the idea in the first place. Both men were constantly looking for new ways to expand their collections, had sponsored previous collecting expeditions, and were eager to gain access to additional Spanish American specimens. Houstoun spent time with them upon his return to England after the sinking of the *Assiento*. He attended Royal Society meetings that autumn as the guest of Miller and likely visited Sloane at his Chelsea estate. Whatever the genesis of the idea, both men strongly encouraged it. They helped Houstoun secure the patrons and permissions he needed to set off on his expedition, and they continued to promote the endeavor after his departure. Sloane wrote to remind patrons of their commitments when they were slow to send the money they had promised. Miller functioned as a clearinghouse for seeds and plants sent by Houstoun, simplifying the process of sharing the expedition's spoils with its patrons. Houstoun knew that if seedlings arrived in poor health or if seeds were difficult to coax into bloom in Britain's climate, Miller would be up to the challenge.[27]

Houstoun's promise to obtain in Spanish America all the "usefull Plants ... which are wanting in our American Colonies" appealed to botanists and gardeners as well as the founders of Georgia.[28] For botanists and gardeners, Houstoun's proposal offered the chance to acquire rare or previously unknown plants native to Spanish America. Elite and, increasingly, middling gardeners vied to acquire new plants brought to Britain on ships returning from Asia, Africa, and the Americas. Subscribing to expeditions such as Houstoun's offered one of the best ways for avid gardeners to add rare plants to their collections. Houstoun's patrons included some of the most enthusiastic British collectors of plants and other natural curiosities. Their sponsorship of the former slave ship surgeon's expedition was part of a larger pattern of subscribing to natural history texts and sponsoring collecting trips. Houstoun's patrons, for example, also supported the natural historical collecting efforts of Mark Catesby and John Bartram in British North America.[29] For the trustees of Georgia, Houstoun's expedition offered the promise of introducing valuable botanical commodities from the Spanish Empire into cultivation in the southernmost mainland British colony. Such natural commodities might help the nascent colony establish an economy independent of slave labor, as its founders intended. Houstoun's eight patrons collectively pledged to provide him with an annual salary of £200. Although the Georgia trustees only contributed £15 toward Houstoun's annual salary, they largely directed his movements. In their eyes, at least, the expedition's primary objective was to acquire plants for their new colony.[30]

Promotional materials for the colony boasted that when Georgia was "well-peopled and rightly cultivated," it would supply Britain with the "raw Silk, Wine, Oil, Dyes, Drugs, and many other materials for manufactures," which Britons traditionally imported "from Southern countries." If these "Southern" commodities could be grown in the southernmost British colony to be planted on the American mainland, Georgia would be set on secure economic footing and the empire's balance of trade would improve. It fell to Houstoun to collect the seeds, roots, and cuttings necessary to make this happen. As one trustee described the naturalist's role, he was intended to be a "collector of drugs and plants of use to be gathered from other countries and planted in Georgia."[31]

The trustees' unique vision for the province made them particularly keen to introduce valuable foreign plants into their province. As historian Mart Stewart has argued, the trustees of Georgia sought to establish a new sort of colony, premised on both philanthropy and natural inquiry. Georgia was intended to provide a haven for Britain's poor and a buffer against Spanish Florida. To achieve both goals, the trustees believed that they needed to establish a colony radically different from South Carolina, its nearest neighbor. In place of large plantations worked by enslaved Africans, the trustees envisioned Georgia as a colony composed of small farms worked by free or indentured Europeans. Such an arrangement would safeguard opportunity for Britain's poor and ensure the presence of a large white population for defense. In order for such a colony to succeed, its economy would need to be as different from that of South Carolina's as the trustees intended its social structure to be. Small farms could not hope to profitably compete with Carolinian plantations in the production of rice or indigo. Instead, Georgia would need to export valuable natural commodities such as silk, wine, drugs, and dyes. For this, the trustees put their faith in natural history.[32]

The trustees of Georgia embraced the idea that natural history and natural philosophy could be harnessed to solve practical problems, including those of establishing a new colony. Their enthusiasm for natural knowledge partially can be gauged by the fact that a third of their original members were also fellows of the Royal Society. The trustees' faith in the utilitarian potential of natural knowledge reflected the broader popularity of such ideas in eighteenth-century Britain following the publication of Sir Isaac Newton's *Principia*. In particular, the trustees emphasized the utility of botanic gardens and economy botany to colonial development. Plans for the colony called for a ten-acre garden in Savannah to serve as both an experimental station to test potential natural commodities and a nursery to provide colonists with those plants that succeeded. The trustees agreed to sponsor Houstoun's expedition just six weeks before the first group of Georgia colonists departed England. Houstoun's appointment as the colony's botanist and traveling naturalist therefore needs to be understood as part of the initial efforts to establish a new colony grounded in both science and philanthropy. Ironically, the trustees planned to exploit the routes of the slave trade to acquire the plants necessary to establish a colony free of slavery.[33]

The trustees of Georgia and Houstoun's other patrons provided more than just financial support. They drew upon their connections in the overlapping worlds of British commerce, empire, and natural history to secure the permissions and introductions that the naturalist's plans required. In his agreement with the trustees of Georgia, Houstoun "beg[ged] of the Hon'ble Trustees, to procure of Sir John Eyles and others of the Directors of the Hon'ble South Sea Company Letters of Recommendation to their respective Factors." His request hinged on the knowledge that the men involved in the philanthropic project to establish Georgia had extensive connections among the wealthy merchants on the South Sea Company's Court of Directors. The successful West Indian merchant George Heathcote, for example, was a member of the South Sea Company's Court of Directors, a fellow of the Royal Society, and the treasurer of the trustees of Georgia. Through his wife, Maria Eyles Heathcote, he was related Sir John Eyles, subgovernor of the South Sea Company's Court of Directors. Sloane, who had proposed Heathcote for membership in the Royal Society in 1729, and other supporters of Houstoun's expedition similarly enjoyed close ties to the South Sea Company's directors. Through these connections Houstoun was able to secure the cooperation of the South Sea Company and letters of introduction to Spanish American officials.[34]

Although the South Sea Company did not contribute a penny toward Houstoun's expedition, it was in many ways his most important patron. The company instructed its agents in Jamaica to give Houstoun (and, later, Millar) free passage on its ships. It provided letters of introduction to its factors in Spanish America and instructed its employees to assist the naturalists in any way they could. For the most part, servants of the South Sea Company did so. Company agents cared for the naturalists' plants, intervened on their behalf with Spanish officials, leveraged their local connections to assist in the collection of specimens, and pledged to forward any seeds or plants they received after the naturalists departed.[35]

The company's willingness to assist Houstoun and Millar in violation of the *asiento* agreement cannot be explained solely by the personal connections of its directors with Houstoun's patrons. The South Sea Company's decision came during a period when the *asiento* relationship was already disintegrating. By the 1730s company officials were increasingly frustrated with Spain's restrictive interpretation of the trade's terms, harassment of the company's personnel, and seizures of its goods and vessels. These tensions contributed to the outbreak of war between the two empires in 1739. The deteriorating relationship between the company and Spain created an environment in which the company's directors likely felt they had little left to lose by assisting Houstoun and Millar, while the British Empire and individual patrons potentially had much to gain.³⁶

"EXCEPT IT BE BY A PERSON FIXED THER[E]"

Although the plans to botanize under the cover of the slave trade were developed by Houstoun, he died just months into the expedition. Houstoun's patrons hired Robert Millar as his replacement and instructed him to follow the plans developed by the slaving surgeon. Yet key differences in Houstoun's and Millar's approaches to the myriad challenges of bioprospecting in a foreign territory had significant impacts on the ultimate outcome. Millar's many frustrations as a biopirate indicate that access in the form of ready transportation and local connections was not sufficient to succeed as a biopirate in Spanish America.

In late October 1732 Houstoun boarded the *Amelia* bound for Jamaica by way of Madeira. Madeira, a Portuguese island set in the Atlantic Ocean approximately 600 miles off the Iberian Peninsula, was renowned for the wine that shared its name. Madeira accounted for nearly all the wine imported into British America in the 1730s. Houstoun spent his time on the island studying the local techniques of wine production and gathered enough cuttings of local grape vines to fill two large tubs. These, he proudly reported, thrived under his care during the subsequent voyage to Jamaica. When he arrived in Kingston on December 20, the vines were already budding and "put[ting] out Shoots of an inch or two long." The naturalist entrusted vine cuttings to the South Sea Company's agent, Pratter, who promised to plant them in his garden on the outskirts of town. These cuttings were destined for Georgia, to see if the colony could produce wines to rival those of Madeira.³⁷

Garden space was not the only favor Houstoun requested when he called on Pratter the morning after his arrival in Jamaica. Pratter's assistance would determine whether the naturalist gained access to Spanish America. As the South Sea Company's representative in Jamaica, Pratter decided when and where the company's slaving vessels would sail and who would be allowed on board them when they did. Houstoun reported with some evident relief that Pratter had agreed to arrange his passage to Cartagena on one of the company's vessels that was scheduled to sail in a few days.³⁸

Houstoun reached the South Sea Company's trading factory in Cartagena on January 3, 1733. He reported that he was "very well received at the Factory on account of one Gentleman who is my Relation and Some former acquaintance I had of the rest." Houstoun enlisted the assistance of his kinsman, the factory's surgeon James Houstoun, and his former colleagues in his quest to procure seeds of valuable medicinal plants indigenous to the area near Cartagena. Houstoun described or collected at least twenty-five different plants in the region. In addition to collecting and describing new plants in Cartagena, Houstoun hoped to locate seeds of well-known and valuable medicaments imported from the Spanish Empire. He believed that *Ipecacuanha* could be found in Mompox, about a week's journey inland from the port city. He anticipated that the "extreamly severe" local governor was unlikely to allow him to travel there to search for the plant. Instead, the naturalist focused on finding alternative means of procuring the seeds. Houstoun convinced a Spanish gentleman who was about to set out for the region to send him the plant. Not one to leave things to chance, he also wrote to three individuals who resided near Mompox to ask that they send *Ipecacuanha* seeds or seedlings to him at the Cartagena factory. Houstoun confessed to his British patrons that the other medicinal plants he hoped to obtain while in Cartagena might prove more difficult. The naturalist promised to "use my utmost endeavours to get the Seeds of the Trees that produce the Balsam[s] called Capivi and of Tolu." Given that these trees could only be found "still further up the Country," they would be "Consequently harder to be come at."[39]

Houstoun did not live long enough to learn whether it would be possible to procure these plants. While collecting specimens in Cartagena in early 1733, he fell dangerously ill with a fever. Although he made a temporary recovery that allowed him to return to Jamaica, his improvement was fleeting. On August 14, 1733, "after a long and severe Illness," Houstoun died, with over two years remaining in the planned three-year expedition.[40] To take his place, Sloane recruited the surgeon Robert Millar, who had previously collected on Sloane's behalf in the Levant. Houstoun's sponsors agreed that Millar would continue the expedition Houstoun had planned and under the terms the former slave ship surgeon had negotiated.[41]

With his appointment as Houstoun's successor, Millar inherited a difficult task. Unlike Houstoun, Millar had no experience with the *asiento*, the slave trade, the South Sea Company, or Spanish America. Perhaps this explains his different approach to the challenge of bioprospecting in a foreign territory. Houstoun had argued that the assistance of the South Sea Company would be crucial and that only the pretense of being engaged in the British slave trade to Spanish America would provide the access necessary to bioprospect in the region. Although Millar continued to rely upon the South Sea Company's vessels for transportation to Spanish America and its personnel for occasional assistance, he disregarded Houstoun's advice to do so under the pretext that he was engaged in the slave trade. Instead, Millar put his faith in the reputation of the Royal Society of London and the deception that he was collecting on its behalf. Although such a strategy worked in some regions, in others Millar received only "Hardship and Cruel usage" for his troubles.[42]

A few months after his appointment as Houstoun's successor, Millar embarked on board the *St. Thomas* bound for Jamaica. Like Houstoun, Millar's first stop upon arriving on the island was the office of the South Sea Company's agents. There he found Pratter happy to help him secure passage to Spanish America. The agent "Immediately G[a]ve me Liberty to go Passenger to any Place on the Continent Where we had factories." Millar chose the linked factories of Portobelo and Panama as his first destination.[43] After a brief stay in Portobelo, Millar headed west to the South Sea Company's factory in Panama. The valuable drugs cinchona and balsam fern were exported from the Pacific port city. Millar "made a Particular Inquiry into the Trees wch yield's" those medicaments. From local informants Millar learned that the balsam fern medicament was "falsly Called So" because most of the drug exported from Panama under that name came from trees growing in the mountains of Nicaragua, not from ferns. Millar's local contacts, however, were not able to help him procure specimens of the medicinal plants. The naturalist declared that he had "used the utmost of My Endeavour to the Purchasing them and to perswade the Gentlemen of the factories to use thers." Neither Millar nor the South Sea Company factors could acquire the desired plants. After collecting what seeds and plants he could find, Millar returned to Portobelo. With the onset of the rainy season and with three South Sea Company factories still to visit, Millar decided it was time to return to Jamaica to await a ship to take him to Cartagena.[44]

In Cartagena Millar was granted extraordinary latitude to travel and to collect specimens thanks to a cooperative local official. The naturalist waited upon the governor of New Granada upon arriving in February 1735 and presented a letter of introduction that his patron, Robert James Petre, Baron Petre, had obtained on his behalf from Cristóbal Gregorio Portocarrero y Funes de Villalpando, the fifth conde de Montijo, Spanish ambassador to London, and member of the Royal Society of London. On Montijo's recommendation, the governor of New Granada granted Millar permission to collect for a month.[45]

Millar encountered three of the medicaments he most desired during his month-long collecting trip: ipecacuanha, balsam capivi, and balsam of Tolu. By the 1730s the emetic ipecacuanha had become so popular among British medical men that "it has almost justled all other Emetics out of Use." Despite its popularity, there was no British source for the plant, so it would have made a valuable addition to the Trustees' Garden in Savannah. Millar was unable to locate any *Ipecacuanha* seeds because he visited Cartagena during the winter. In the lieu of seeds, he gathered over a hundred seedlings of the plant. The naturalist had better luck locating seeds for the trees that produced the balsams of Tolu and capivi. Like ipecacuanha, both balsams were esteemed in Britain for their medicinal properties. The balsam capivi was prized as a cure for gonorrhea and for wounds, while the balsam of Tolu was considered an "excellent pectoral Balsam, of great Service in Affections of the Lungs, as Coughs, Asthmas, [and] Consumptions." Millar also carefully noted the manner by which locals extracted the balsams from the trees. Throughout his month of liberty, the naturalist gathered seeds of "many other Trees & Plants of less note that grew therabout." Many of these plants "of less note" turned out to be previously unknown to European botany. Nearly half of all the plants Millar is credited with introducing into Britain over the course of his career were gathered during his month in Cartagena.[46]

During Millar's first two trips along the routes of the *asiento*, he enjoyed access to collect in the regions surrounding South Sea Company factories. His experiences in Veracruz, however, were quite different. Millar arrived in Veracruz hopeful that he would be allowed to travel inland in search of cochineal, jalap, and "Several other usefull Drugs." Instead, he was not even allowed to disembark. On the orders of Adm. Manuel Lopéz Pintado, Millar was confined to the vessel on which he arrived. The naturalist turned to the South Sea Company factors in Veracruz for help, trusting that they would have useful friends among local officials. Initially the factors' influence seemed to do the trick. The local governor ordered that Millar be allowed to come ashore. Admiral Pintado, however, remained determined to prevent the naturalist from bioprospecting in New Spain. When he discovered that the governor had countermanded his order, the admiral seized Millar's trunks, including his clothing, seeds, seedlings, descriptions of Campeche plants, and other papers. Once again Millar put his faith in the South Sea Company factors' ability to intercede on his behalf. The factors accompanied Millar when he went to the custom house to demand the return of his belongings. Pintado remained resolute, declaring that he would not return Millar's trunks until he boarded a Spanish man-of-war bound for Havana. Ultimately, Millar was told nothing could be done for him and he reluctantly boarded a Spanish vessel bound for Cuba. During the voyage to Havana, the vessel's captain "acquainted [Millar] of the Rigour of the Governor of *Havannah* & told me that the only usage I could expect from him was to be sent to Old Spain." Luckily for Millar, as the Spanish ship came within sight of Cuba, it met a British vessel leaving the harbor. The sympathetic Spanish captain told Millar that the "Best & kindness thing he could do" for him was to put him on board the British ship. Millar and his belongings were consequently transferred to the *Constant*, bound for England.[47]

Millar made a second attempt to collect in Veracruz, but it was as unsuccessful as the first. For his second trip to Veracruz in 1738 Millar secured another letter of introduction from the conde de Montijo, this time to Juan Antonio de Vizarrón y Eguiarreta, the archbishop of Mexico and viceroy of New Spain.[48] Montijo's letter erroneously stated that Millar had been sent to Spanish America on behalf of the Royal Society in order to discover "vegetables and other curiosities." Citing his own membership in the Royal Society as the reason, Montijo requested that the viceroy offer Millar his protection and favor. Yet Millar's expedition was not sponsored by the Royal Society. The naturalist's patrons included a few fellows of the Royal Society, but the organization itself was not involved in his expedition in any way. By framing Millar's mission as one sponsored by the Royal Society and undertaken in the name of natural knowledge rather than sponsored by, among others, the trustees of Georgia in the name of British imperial ambitions, Montijo's letter presented Millar's expedition as relatively benign. In the end, however, even Montijo's misleading description of the expedition was of little use.[49]

Although the local governor and other Spanish officials received Millar "with a great show of Civility," the viceroy was as unyielding as Admiral Pintado had been. Vizarrón y Eguiarreta forbade Millar from traveling in New Spain, explaining that his instructions "were such, that he Could not Give that Liberty to any Person" without the monarch's express permission. Further, he ordered that Millar be confined to the city walls without access to his baggage until he returned to Jamaica. As the naturalist later recounted, "The little Hardships & Cruel usage I then underwent was somewhat Severe." For the next two months, the naturalist waited for a ship bound for British America. During this time, he was denied the use of his books, linen, clothing, and "even a few Necessaries for [his] Support."[50]

This second failed expedition to Veracruz convinced Millar to retire from the life of a biopirate. In his final letter to the trustees of Georgia, he explained that he thought "it will be to no manner of Purpose, my attempting anything further, in this way, having already mett with so many Rubs & Dissapointments." He therefore resigned from his post but promised that on his return voyage he would deliver to Georgia the plants he obtained in Cartagena and nursed back to health.[51]

Reflecting on his "Rubs & Dissapointments," Millar concluded that a British collector's success in Spanish America hinged on the South Sea Company. The naturalist blamed his hardships in Veracruz on the weakness of the company's local influence. Moreover, he declared that all of his experiences led him to believe that "ther[e] is no Possibility of doing any thing in that Kingdom, except it be by a Person fixed ther in some Station in the Company Service." In other words, after four years bioprospecting in Spanish America, Millar resolved that collecting in New Spain was an impossible task for a traveling naturalist. The only way to succeed as a British biopirate in Spanish America was to be "fixed" in Spanish territories as a resident factor engaged in the *asiento* slave trade on the South Sea's Company behalf.[52]

THE PROCEEDS OF BIOPIRACY

The question of whether Houstoun's and Millar's biopiracy expeditions were successful was ultimately a matter of perspective. For the Georgia trustees, the undertaking had largely been a failure. Their garden in Savannah received precious little in return for their six years of support. More broadly, Houstoun and Millar failed to acquire natural commodities that would have most benefited British imperial expansion and trade, such as cochineal or cinchona. Judged by the expeditions' impact on British natural history and gardening, however, they were much more successful. Plants acquired by Houstoun and Millar lived on in British gardens, herbariums, and scientific texts.

Reflecting on the Georgia trustees' patronage of Houstoun and Millar, one trustee concluded that they had "seen no fruits of our expense, but a disappointment of our expectations." The trustees' expectation that Houstoun and Millar would gather many natural commodities on behalf of Georgia was never met. The only plant that we can say with certainty reached the colony were the tubs of grape vines that Houstoun collected in Madeira on his outward voyage in 1732.[53] During the four years Millar collected along the routes of the *asiento*, he repeatedly assured the trustees of Georgia that he had saved his best seeds for the new colony. It seems unlikely that any of these ever reached Georgia. Millar's agreement with the trustees also instructed him to personally oversee the transplantation of his plants and seeds into the Trustees' Garden in Savannah. The naturalist only made it as far as South Carolina. It is possible that while there he entrusted *Ipecacuanha* and other plants to the trustee's agent in Charles Town. Regardless of whether he did so, the trustees continued to believe they had been ill used by Millar.[54]

There is no ambiguity surrounding the impact of Millar's and Houstoun's collections on British gardens and natural history. Philip Miller received hundreds of seeds acquired by the biopirates along the routes of the *asiento* trade. In 1737 alone, Robert Millar sent seeds for 212 different plants to the Chelsea gardener. Similarly, Sloane received a collection of 347 seeds from Cartagena and Panama during the same year.[55] During the brief period that Houstoun was well enough to collect in Cartagena, he gathered seeds from plants representing at least 25 different genera. These plants were later grown in the Chelsea Physic Garden. Miller noted in his *Gardeners Dictionary*, for example, that he received the *Asphodel* lily from Cartagena, "which have multiplied greatly in the Chelsea Garden."[56] Other expedition patrons, such as Baron Petre and Charles Du Bois, also successfully grew Spanish American and Jamaican plants in their gardens from seeds they received from Houstoun and Millar. The seeds, botanical descriptions, and herbarium specimens Houstoun and Millar acquired in Spanish America continued to shape natural knowledge long after the collectors had passed from the scene.[57]

The impact of Houstoun's and Millar's collecting efforts was amplified by the common practice among eighteenth-century naturalists and gardeners of sharing duplicate seeds and seedlings. Therefore, plants collected by Houstoun and Millar circulated beyond their small group of patrons. For instance, in 1734 Philip Miller sent the British botanist Richard Richardson seeds that Houstoun collected in Cartagena. In return the Chelsea gardener requested seeds for any northern or Welsh plants in Richardson's collection. Other patrons similarly shared the botanical spoils of the expeditions. Petre's splendid collection of American plants in his gardens at Thornton were largely sold off by his widow following his unexpected death from smallpox in 1742. Yet Petre's generosity in sharing seeds and seedlings during his lifetime meant that almost fifty years later, plants grown from seeds at Thornton and gifted to other collectors could be found growing in the Royal Botanic Garden at Kew. These plants included Spanish American species originally collected along the routes of the *asiento* by Houstoun and Millar.[58]

CONCLUSION

The British slave trade to Spanish America provided unparalleled access for South Sea Company employees such as Houstoun to observe, obtain, and smuggle Spanish America's flora and fauna. Their ability to do so rested on the knowledge and labor of local guides, hunters, medical practitioners, and collectors, particularly those of African and Amerindian descent. British naturalists were eager to exploit the openings to Spanish America created by the trade in African captives. The ease with which Houstoun secured patrons to sponsor his biopiracy expedition reflected this eagerness. Houstoun would have understood that his proposal to return to Spanish America to bioprospect under the cover of the slave trade was inherently risky. His actions violated both Spanish imperial regulations and the *asiento* contract. Yet, like smugglers throughout the Atlantic World, he concluded that the potential profits outweighed the risks.

Contemporaries understood that the slave trade could be used "as a lever to pry open the gates of Spanish America to a broader commerce."[59] British naturalists exploited this to pry open access to the legendary riches of the region's natural world. They sought to use the commerce in African captives to facilitate natural historical collecting in Spanish America. In recent years historians of science have emphasized the varied ways that science relied upon and shaped commerce in the early modern period. This literature has highlighted how naturalists utilized commercial communication systems, vessels, trading routes, and systems of credit.[60] Commercial infrastructures facilitated the production and circulation of natural knowledge. As Millar's experiences reveal, these infrastructures did not work equally well in all contexts. In some cases European naturalists could catch a ride and collect wherever commercial vessels took them without the naturalists themselves being directly involved in the trade upon which they relied. Spain's determination to keep foreigners out of its territories and to protect its monopoly on natural commodities meant that such a strategy was unlikely to work in Spanish America. As Millar's difficulties in New Spain illustrate, British natural historical collecting in Spanish America required that the collector be—or plausibly feign to be—engaged in the slave trade itself.

Slave ship surgeons such as Houstoun were not the only mariners to exploit the slave trade to obtain specimens on behalf of eager metropolitan naturalists. The next chapter examines how London silversmith and naturalist Dru Drury recruited mariners to acquire insects on his behalf along the West African coast. Drury relied on the expanding commercial and naval circuits that connected Britain and West Africa in the mid-eighteenth century, especially those of the transatlantic slave trade, in his attempt to acquire one of the largest beetles known in the eighteenth century.

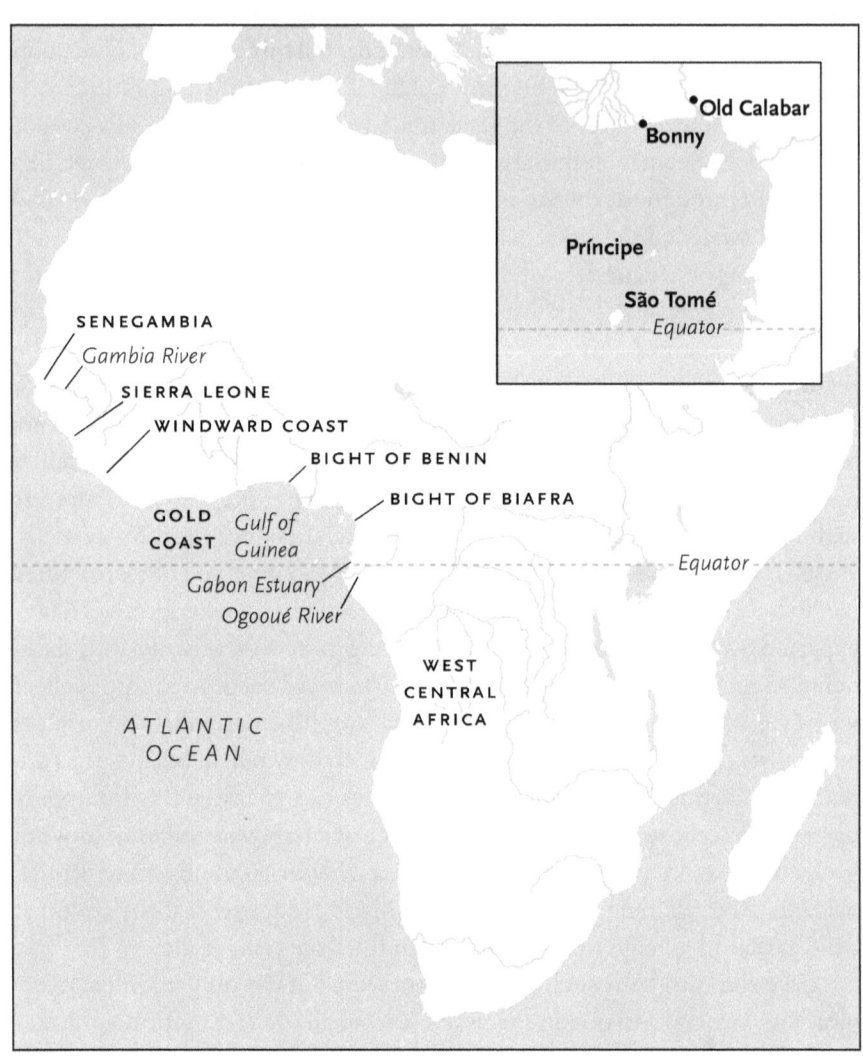

MAP 5.1. Provenance of *Goliathus goliatus* and locations where additional specimens were subsequently sought in West and West Central Africa. The Goliath beetle was found in the Gabon Estuary near the equator, but Drury also tried to obtain specimens from regions farther north, where British mariners more frequently visited.

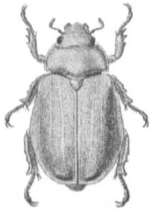

5.

Searching for Goliath

It was the sort of beetle about which collectors dreamed. The Goliath beetle, as it came to be called, measured nearly four inches in length, making it one of the largest known insects. It was distinguished by eye-catching white and pink markings set off against a black thorax (fig. 5.1). A ship captain found it in 1766 floating in the Gabon Estuary, just north of the equator in the Gulf of Guinea. The merchant captain who found the beetle gave it to David Ogilvie, a naval surgeon. Back in London, the British surgeon showed the Goliath beetle to many of the metropole's leading naturalists and collectors. The beetle's size, beauty, and rarity quickly made it an object of desire among naturalists.[1]

Among those most covetous of the Goliath beetle was Dru Drury. The London silversmith was an avid collector of insects and the author of the three-volume entomological text *Illustrations of Natural History* (1770–82). Drury's natural historical museum, at the time of his death in 1804, was estimated to contain 11,000 specimens. Drury relied on routes of British commerce and colonialism to assemble a global collection of natural historical specimens without leaving Britain. Drury's acquisitions were single-minded. As he explained to a perspective collaborator, "my taste in the subject is confined intirely to Insects. All the other parts of Natural History put together does not afford me a quant[ity] of delight as this single class." Drury's

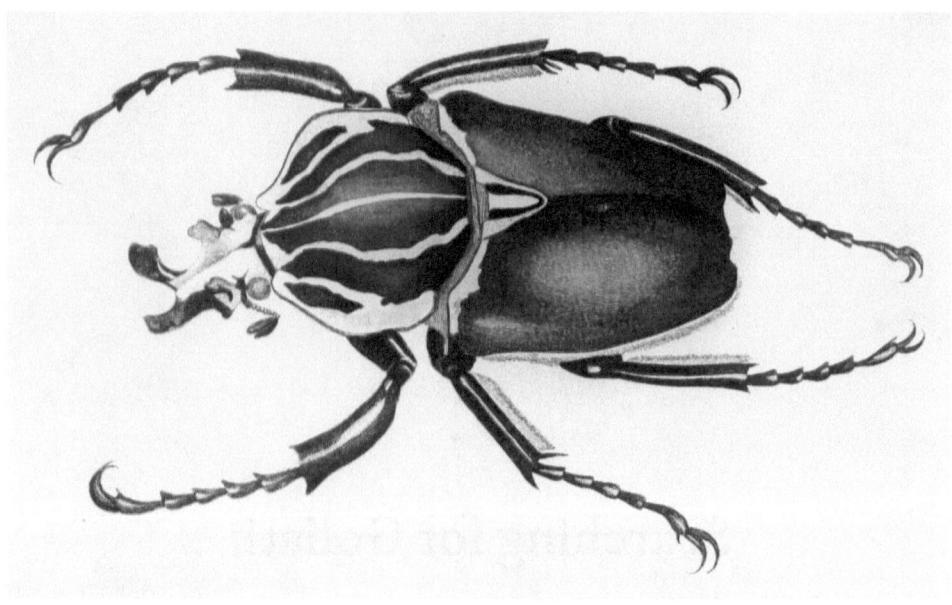

FIGURE 5.1. Goliath beetle (*Goliathus goliatus*). Dru Drury's *Illustrations of Natural History* (1770) featured the extraordinary Goliath beetle that was found floating in the Gabon Estuary in 1766. The Goliath inspired Drury to redouble his efforts to obtain African insects by recruiting mariners as paid specimen collectors. Dru Drury, *Illustrations of Natural History*, London, 1770, Vol. 1, Tab. 31, RB 152307, Huntington Library, San Marino, California.

prosperous business as a silversmith enabled him to indulge his "taste" for entomology. He purchased specimens from other collectors, took out advertisements in foreign newspapers to announce his willingness to pay for specimens, and recruited potential collectors from among Britain's mariners, colonists, and other travelers. After Drury viewed Ogilvie's Goliath beetle in 1766, obtaining one for his own collection became something of an obsession.[2]

Drury had attempted to acquire African insects before the discovery of the Goliath beetle but with limited results. The extraordinary beetle spurred him to redouble his efforts. Drury turned to the commercial and naval circuits that connected Britain and coastal Africa in the mid-eighteenth century in order to do so. The naturalist's confidence that he could obtain a Goliath beetle by means of the mariners who plied Britain's routes to West and West Central Africa reflected the expanded scale of Britain's commercial and colonial engagement with the region in the mid-eighteenth century. In particular, it reflected Britain's expanded participation in the transatlantic slave trade.

During the half century before the Goliath beetle was fished out of the Gabon Estuary, British presence along the African coast significantly increased. British trade in both natural commodities and human cargo steadily rose over this period, albeit with temporary wartime interruptions. More than twice as many enslaved Africans were shipped on British vessels in the third quarter of the eighteenth century than during the century's first quarter.[3] Britain's victory in the Seven Years' War brought new territories into its imperial orbit, including its first in West Africa, the short-lived British colony of Senegambia.[4]

Despite British efforts to establish a colony in Senegambia, British presence in West and West Central Africa during the mid-eighteenth century remained largely confined to the coast. The decline and eventual dissolution of the Royal African Company had opened the African trade to all British merchants in the early eighteenth century. Yet the company's argument that its forts were essential to the success of British trade in West Africa remained widely held. Parliament therefore created the Company of Merchants Trading to Africa in 1750 to assume responsibility for the upkeep and management of forts, factories, and settlements that had previously belonged to the Royal African Company. Unlike its predecessors, the Company of Merchants was not a trading company. Instead, its sole charge was to maintain and manage British forts using funds provided by Parliament. To assist the company in maintaining Britain's foothold in West Africa, the 1750 act also established a system of annual naval patrols to the African coast. As historian Joshua Newton has argued, the Royal Navy's annual patrols represented "a major facet of state participation and collaboration with the slave trade." Drury recruited potential collectors from mariners bound for West and West Central Africa on both merchant ships and royal naval vessels on the annual African patrol.[5]

For metropolitan naturalists such as Drury, Britain's expanding presence along the West and West Central African coasts during the mid-eighteenth century meant new opportunities to acquire specimens through the same circuits of the slave trade. It meant more British ships visited West and West Central African shores each year and therefore more British mariners who potentially could be recruited to act as specimen-hunters along the African coast. To encourage mariners to obtain insect specimens, Drury provided collecting supplies, images of what he desired, directions, and, most importantly, promises to pay for each specimen he received. By providing a reliable and ready market for specimens, he converted insects into commodities to be extracted.

Drury believed that the British slave trade would not only enable him to acquire new African specimens but also allow him to obtain a Goliath beetle in particular. He admitted that the specimen's discovery floating in a river meant that he could not be certain that its natural habitat was close to where it had been found. Yet he speculated that the specimen's excellent state of preservation indicated that it had not traveled far. Unfortunately for Drury, he was wrong on this point. The species of Goliath beetle acquired by Ogilvie (*Goliathus goliatus*) is indigenous to Central Africa, far from the coast where Drury's maritime collectors visited. This likely explains Drury's failure to obtain a specimen for his own museum.[6] Yet his many efforts to do so allow us to trace the techniques of collecting natural historical specimens by means of the slave trade in the mid-eighteenth century. The silversmith's correspondence, account books, publications, and museum catalog allow us to examine how an avid metropolitan collector might utilize British commercial and naval circuits to Africa in the pursuit of a particularly desirable specimen. In the search for the Goliath beetle, the naturalist repeatedly articulated ways that collecting specimens could be integrated into the business of enslaving.

A BEETLE IN A HAYSTACK

Capt. Nonus Parke commanded at least five slaving voyages to the Bight of Biafra between 1759 and 1772. In April 1769 Parke was preparing to depart London for Liverpool, where he would take command of the *Hector*, a slaving vessel bound for Calabar. Before leaving town the captain called upon his London friends to say his goodbyes. When he visited Dru Drury, the silversmith presented him with a stack of engraved prints of the Goliath beetle. Drury likely explained that the rare and beautiful insect had been collected from the Gabon Estuary and that he greatly desired to acquire one for his museum. The naturalist declared that he knew "of no possible method for doing it but by applying to Persons going there." Most Britons bound for Gabon were engaged in the commerce in African captives or in woods, such as dyewoods. As Parke was going to Liverpool "from whence many Ships go to Gaboon," Drury requested that the captain distribute copies of the Goliath print to the chief mates on vessels sailing to the region (map 5.1). If any mate managed to use the print to locate the Goliath beetle, Drury would have gladly purchased it for his museum.[7]

In the years following the discovery of the Goliath beetle, Drury frequently applied to individuals traveling to West and West Central Africa, especially those like Parke going to Old Calabar and regions farther south, near the Gabon Estuary. As Drury explained to Parke, he believed the best way to obtain a specimen of the Goliath beetle was by means of the mariners who traveled to where the original specimen had been found. Drury therefore recruited potential collectors from among British mariners bound for West and West Central Africa. He provided them with prints of the Goliath beetle and with the supplies that they would need to gather such insects.

The Goliath print that Drury distributed was a subject of controversy within the British natural historical community. Drury included the print as plate 31 in the first volume of his *Illustrations of Natural History* (1770). Elsewhere in the text he always acknowledged the owners of any specimens that were not part of his personal collection. However, he failed to do so for the Goliath beetle. This omission was all the more controversial because of the means by which Drury obtained the plate depicting the Goliath beetle. At some point before August 1767, Ogilvie sold the Goliath specimen to Dr. William Hunter, a celebrated anatomist and wealthy physician. Hunter loaned the specimen to naturalist Emanuel Mendes da Costa so that da Costa could have an engraving of it made to include in his planned *Gleanings of Natural History*. Da Costa engaged artist and entomologist Moses Harris to produce the engraving. However, embezzlement landed da Costa in jail in November 1767, and most of the engravings and plates for his planned text were auctioned off to pay his prison costs. One of the few plates to survive the auction block was that of the Goliath beetle. In February 1768 da Costa quietly sold the plate to Drury for three guineas, without Hunter's permission or knowledge. By including the Goliath plate in *Illustrations* without reference to Hunter, Drury implied that he was the owner of the rare and beautiful specimen. Hunter was furious. He believed da Costa and Drury guilty of deception, as he made abundantly clear in a biting letter to da Costa.[8]

Contemporaries and most historians of the episode have focused on the means through which Drury obtained the Goliath plate and the impropriety of including it in his *Illustrations*. What has received little attention is the other use to which Drury put the plate. Drury had prints made from it almost immediately and hired Harris to hand-color them.[9] Some of these prints were later included in the first volume of *Illustrations*. Yet others, like those entrusted to Parke, were intended as a technology of collection. Drury gave the Goliath print to mariners bound for West and West Central Africa in order to assist them in locating additional specimens of the beetle. The prints were a tool for finding a beetle in a proverbial haystack.

At least five captains bound for West or West Central Africa between 1768 and 1771 received copies of the Goliath print from Drury. They were in addition to the Liverpudlian ship's mates who may have received copies from Parke in 1769. The five captains who received the Goliath print included three slaving captains, a royal naval captain, and a merchant captain and rice planter returning to West Florida by way of West Central Africa. All five recipients were bound for the Bight of Biafra, São Tomé, or Príncipe.[10] Drury was uncertain whether the beetle could be found at slaving ports in the Bight of Biafra such as Calabar and Bonny, where British ships more frequently called but were well north of the River Gabon. Undeterred, the naturalist optimistically declared that, with a little investigation, his correspondents could find his beetle.[11]

The first to receive a Goliath print was Capt. Thomas Male, commander of HMS *Hound* assigned to the annual naval patrol of the West African coast. Male's mission was to help Britain retain its foothold in West Africa, especially its slaving forts under the management of the Company of Merchants. In October 1768 Drury "beg[ged] . . . the favour" of Male that if the captain should anchor anywhere on the African coast near the equatorial line, he would use the Goliath print to search for specimens of the beetle. The naturalist instructed Male to "shew the print to some of the Natives there & enquire if they know such an Insect by the representation."[12]

Drury made similar requests to slaving captains visiting the region. In December 1768 the silversmith asked Capt. Thomas Williams of the slave ship *Meredith* to show copies of the Goliath print to the Africans he met on a slaving voyage to Sierra Leone. "Enquire if they know such a Creature or can procure any such," the naturalist instructed. "If they can I will intreat you to bring me some (a dozen if it is possible)," Drury continued. The next year, the naturalist continued his search for the Goliath beetle when he gave Parke copies of the Goliath print for mates on board Liverpool-based vessels. The vessels on which these mates sailed would have been engaged in either the trade in natural commodities or the slave trade. Although these initial efforts yielded no Goliath beetles, Drury persisted. The naturalist reminded Captain Parke to look for the Goliath beetle two years later when the captain was preparing for his next slaving voyage to Calabar. "But above all things let me intreat you to procure as many as you possibly can of the Beetles represented in the print I sent you [when] you went but last time," the naturalist pleaded.[13] At its simplest, each recipient of a Goliath print received the same instruction: show it to all the Africans you meet along the coast and procure as many Goliath beetles as possible.

Drury's plan for procuring the Goliath beetle rested on the belief that Africans would, as he told Captain Male, "know such an Insect by the representation." It hinged on a faith in local natural knowledge and in the power of visual, rather than textual, representations of nature. Rather than ask mariners to search for the beetle themselves or to describe what they were looking for, Drury instructed them to circulate the Goliath print among the Africans they met. The naturalist assumed that local informants would know the insect from its visual representation and be able to locate additional specimens of it. He implicitly acknowledged European naturalists' reliance on Africans as specimen collectors, hunters, guides, and informants.[14]

Drury's use of the Goliath print as a technology of collection was more than a pragmatic solution to the problems of communicating across cultures and collecting at a distance. It also reflected Drury's belief in the inherent superiority of visual representations of natural specimens over their textual descriptions. The very title of Drury's three-volume *Illustrations of Natural History* highlighted the importance that the naturalist placed on images. Drury declared that *Illustrations*' most significant contribution was the high quality of its engravings, which were drawn, etched, and hand-colored by the talented Moses Harris. In the preface to the first volume of *Illustrations*, Drury laid out his reasons for preferring engravings over verbal descriptions. The naturalist argued that verbal descriptions of specimens required readers to recall all that they already knew, invoking "all our powers of conception, to our assistance" in order to understand "what is intended to be described." Even a description written by a talented author, Drury asserted, often "puzzles and confounds the mind." In contrast, the naturalist argued that well-executed engravings represented to the mind the natural object immediately upon viewing the image, without any need for viewers to call upon previous knowledge.[15]

Drury's preference for visual representations of nature was shared by most eighteenth-century naturalists. As historian Daniela Bleichmar has argued, a collective visual epistemology was central to the practice of eighteenth-century natural history. Naturalists believed that "vision constitutes the best method for investigating nature and that images provide the preferred means of transmitting this knowledge." They understood the trained eye as the essential tool of the naturalist and expert visual skill as the basis of the practice of natural history. Shared methods for understanding, analyzing, and producing natural historical images allowed individuals separated by thousands of miles to collectively see and study nature.[16]

Illustrations' 150 engravings would have allowed naturalists to study the many rare insects in Drury's collection without having to travel to London to observe them in person. Yet some of these images circulated through the routes of the slave trade before naturalists could view them in a published text. Mariners such as Parke received hand-colored images from *Illustrations* two years before the book's first volume went on sale. These images included the Goliath plate as well as representations of other African insects Drury hoped to acquire. Natural history images—both published and unpublished—typically brought specimens from the field into naturalists' distant studies. Drury essentially reversed this by bringing images of specimens from the study into the field as a visual desiderata.[17]

Drury's use of prints parallels in some ways how early modern naturalists employed published natural histories within the commercial and gift economies of specimen exchange. As historian Dániel Margócsy has argued, published natural historical texts functioned like "'mail-order catalogues,' for naturalists who wanted to specify which exact species they wanted to order from their providers." Naturalists who wished to purchase or barter for specimens from a distant correspondent needed to ensure that they were talking about the same species. Yet because there was no agreed-upon universal classification system, one specimen was known by many names. Naturalists therefore turned to illustrated natural historical texts in order to communicate their desires. They trusted that while they might have different local names for plants and animals, they had many of the same books on their shelves. Their requests for specimens were couched in citations to specific images depicted in widely owned texts. Published engravings of specimens became the preferred way for naturalists to communicate with suppliers of specimens. Drury's distribution of Goliath prints similarly used visual depictions of species to communicate his desires at a distance, although he did so to British mariners and African informants rather than to European naturalists or merchants of *naturalia*.[18]

Drury's strategies for obtaining a Goliath specimen for his own collection extended beyond the distribution of prints. He also provided mariners traveling to West and West Central Africa collecting supplies and advice on how to use them. For example, alongside the copies of the Goliath print that Captain Williams of the slave ship *Meredith* received in December 1768 was a large box of "Instruments of various Sorts for taking Insects." The collecting supplies inside the box included different types of nets for capturing flying insects, two pairs of forceps, and a larger batfowling net such as that depicted in figure 5.2. Drury included two pincushions, similar to the one suspended from the naturalist's waistcoat in the figure. He informed Williams that the pins he sent ensured that "you might not be at a Loss for means to fill the Large Box." Piercing the thorax of winged insects such as butterflies with

FIGURE 5.2. Eighteenth-century insect collectors. The frontispiece to Moses Harris's *The Aurelian* depicts two eighteenth-century insect collectors in the field. Their collecting equipment, including batfowling nets, collecting boxes, and pincushions (suspended from the naturalist's waistcoat), is similar to the equipment Dru Drury provided mariners in his boxes of collecting supplies. Moses Harris, *The Aurelian: A Natural History of English Moths and Butterflies*, London, 1766, RB 145975, Huntington Library, San Marino, California.

a pin and attaching them to a cork-lined box helped to preserve delicate specimens in the field. Pins were also used to pack specimens for transport and to display them in cabinets. Drury gave the slaving captain a small box containing two butterflies that had wings with ragged edges to illustrate what *not* to collect. The naturalist explained that such damaged specimens should be thrown away and that only "compleat & perfect" ones should be retained. Drury's rules for collecting were "more particularly described" by the final item in the box, a copy of the naturalist's "Directions for Collecting Insects in Foreign Countries."[19]

Drury sent each newly recruited collector his instructions for how to find, kill, preserve, and pack insect specimens. At some point before 1772 the naturalist had his "Directions" printed, thus saving himself from having to recopy the same instructions for each of his new recruits. The printed version included a blank allowing the naturalist to quickly tailor his requests to the recipients' intended itinerary: "It is desired, if the ship touches at different places in _____ to collect some at each of them." "Directions for Collecting Insects in Foreign Countries" was primarily designed to be used by maritime collectors. As Drury apologetically explained when sending a copy for a gentleman who promised to collect in Surinam, "The written directions sent in the Box are such as I give the Sailors therefore pray ask an excuse for if I had more time I should write a Sett proper as for the Gentlemans perusal."[20]

Drury's "Directions" advised mariner-collectors that he desired specimens of beetles, "FLIES with transparent wings," grasshoppers, ants, fireflies, and any "other sort" except cockroaches, centipedes, and scorpions, "which are in general so very common as not to be worth bringing." It gave potential collectors detailed, practical advice such as that dung heaps and rotten wood were good places to find beetles, that specimens should not be placed into the packing box "till they are quite dead" lest they "pull and tear" their neighbors, and that while most insects could be killed by holding them close to a fire, butterflies and moths should be killed by squeezing their bodies lest a fire damage their delicate wings. Drury advised that once a packing box containing specimens was filled, it should be sealed with a piece of linen dipped in tar to prevent "small vermin getting in, who will certainly do mischief to the insects withinside." Perhaps most important to Drury's ultimate acquisition of a large collection of insects, "Directions" concluded by promising potential mariner-collectors that Drury would pay sixpence for each insect brought to his home in London.[21]

Drury seems to have furnished British slaving mariners and other potential collectors with a standardized set of supplies. Most eighteenth-century instructions for entomological collecting detailed the different collecting equipment available for purchase. These discussions functioned as a guide to help readers decide in which equipment they wished to invest. Drury's "Directions" omitted such a discussion. Instead, it assumed that readers would all have the same type of nets, collecting boxes, and pins at their disposal. It instructed, for example, that the collector heading to the field take "with him the brass tongs, nets &c., the oval pocketbox, and the pincushion stocked with pins." Drury's consistent use of the definite article indicates that "Directions" was intended to accompany his boxes of supplies and that the contents of those boxes were standardized. The recipient of one of Drury's large boxes could expect to find inside nets, tongs, small collecting boxes for the field, and a pincushion complete with pins of various sizes. Drury's boxes of collecting supplies provided everything a novice would need to gather insects on a journey. They were essentially entomological starter packs.[22]

Drury reassured the mariners who received his boxes that collecting natural historical specimens would not interfere with "the ship's business," including that of the slave trade. To the contrary, he argued that the normal workings of British naval and merchant vessels provided numerous opportunities to obtain insects. For example, the naturalist explained that "a great many insects" were often blown on board by evening breezes when ships lay at anchor. Drury therefore recommended that would-be collectors make a thorough search of their ship each morning, using brass tongs or forceps to gather the insects they found. Drury's conversations with Mr. Hough, an experienced slave ship surgeon, led him to conclude that the surgeon could gather specimens during his next slaving voyage without ever stepping foot ashore. Hough had told the naturalist about canoemen, local Africans who ferried trade goods, supplies, and captive Africans between the vessel and the shore. The deadly surf and lack of harbors along parts of the West African coast meant that in some slaving regions only West African canoemen had the necessary maritime skills to safely move between ship and shore. Drury suggested that Hough employ "the Men you mention'd to me who come to the Ship every morning" as specimen collectors. Drury predicted that for a "trifling reward" they would bring whatever insects they saw "in the road coming to the Ship." The naturalist offered not only to reimburse Hough for the "trifling" cost but also to make him an "adequate return" for his efforts.[23] Similarly, in July 1769 Drury suggested that the slave ship captain Parke integrate collecting into the day-to-day duties of his crew. The naturalist suggested that when members of Parke's crew went ashore each day for wood, water, provisions, and to purchase enslaved Africans, they could obtain insects at the same time. In this manner, Drury "imagine[d] Insects might be procured without any prejudice to the Ships business"—namely, the purchasing of enslaved Africans.[24]

Not one to limit the avenues by which he might acquire new specimens, Drury also counseled slaving mariners to hire Africans as specimen collectors. The naturalist told Parke that he understood the slaving captain's "time may be employed so much about Business as not to permit you to entertain any thoughts about my affairs." Employing members of Parke's crew as collectors offered one solution. If this proved impractical, Drury urged the slaving captain to "speak to some of the black people" about procuring specimens. "This I fancy is a method as easily to be pursued as any other," the naturalist explained, and it was likely to produce a superior collection because "the Blacks may know where to procure some curious Insects that your own people may be unacquainted with." Drury speculated that local Africans hired specifically as specimen collectors were also more likely to gather insects in a diversity of habitats and climates. Drury explained to the slave ship surgeon Hough that while it might be more convenient to hire as collectors the African canoemen who visited the ship each morning, these individuals were likely to only gather what they "casually" saw in the road. A dedicated collector, by contrast, could be instructed to spend a few days at a time searching for specimens and "can allow himself to search in more places than the other Men who cannot be permitted to loiter their time away in the pursuit." Drury's advice that slaving mariners rely on African specimen collectors reflected the role of peoples of African descent throughout the Atlantic World selecting, obtaining, preserving, and transporting natural historical specimens.[25]

Drury's directions, collecting supplies, and prints can be understood both as gifts intended to encourage potential collectors to keep their promises and as an effort to discipline their actions if they did so. Prints such as those of the Goliath beetle delineated what Drury desired, while his directions told them how to find, preserve, and transport specimens using the supplies that he provided. The naturalist's injunction that his correspondents should frequently reread his letters and directions reflects a distant collector's concern about his inability to control the actions of far-flung collaborators.[26] Although such a concern was shared by many collectors, Drury's reliance on mariners may have made these concerns particularly acute. As we will see in the next section, Drury's maritime collectors included carpenters and mates as well as the surgeons and captains who were James Petiver's collectors fifty years earlier. Drury's more diverse group of maritime collectors made it all the more essential that he provide them with clear guidance. Detailed directions, provision of all necessary collecting supplies, and explicit promises of payment for each specimen made it possible for Drury to persuade British slaving mariners to become collectors.

RECRUITING AURELIANS

In May 1767 Drury asked a Mr. Cleland, first mate on the slave ship *Canterbury*, to collect for him during a slaving voyage from London to Calabar and Puerto Rico. To enable him to do so, Drury gave the first mate a small box of collecting supplies. Cleland seems never to have used the supplies that he received. Drury recorded that Cleland brought no specimens home with him when the *Canterbury* returned to London in February 1769. Yet the mate helped Drury make an important addition to his other collection, that of potential collectors. Drury's relationship with Cleland's commanding officer, Capt. Nonus Parke, dates to the *Canterbury*'s return in 1769. In that year Parke presented Drury with two *Scarabeus* beetles he had gathered in Calabar. More importantly, Parke promised to gather additional specimens on his next slaving voyage.[27]

Drury understood that the first step to collecting specimens was to collect the people who might obtain *naturalia* on his behalf.[28] Around the same time that Drury first observed the Goliath beetle, he began to record his collection of potential collectors in a ledger he titled "Account of Boxes with Instruments &c for catching Insects deliver'd to Different Persons &c."[29] The ledger recorded each individual who received one of his boxes of collecting supplies. It also indicated the name of the vessels on which the potential collectors were traveling, where their vessels were bound, and how many boxes of collecting supplies they received.[30] It recorded, for example, that Drury delivered a small box of collecting supplies to Cleland on board the *Canterbury* on May 9, 1767, but that Cleland "Bro't none home" when the vessel returned to London in 1769. The ledger indicates that Drury was willing to make a greater investment in Parke's potential as a collector. In 1769 the slaving captain received not only a large box of collecting supplies but also three additional nets and four extra boxes for transporting specimens.[31]

Drury's "Account of Boxes" tracked the naturalist's efforts over a seven-year period to acquire new entomological specimens by means of individuals such as Cleland and Parke. Its more than 150 entries identified Britons bound for distant ports who each received at least one box of collecting supplies from the naturalist. These boxes were intended, Drury explained, "to set up an Aurelian (for so we call a person conversant in the Study of Insects)."[32] Drury's ledger represented a balance sheet of sorts, reflecting the profits and losses sustained when one naturalist got into the business of setting up Aurelians in the hope of obtaining specimens like the Goliath beetle. It allows us to see who Drury recruited to collect in West and West Central Africa, what they received from him, where they were headed, and what (if anything) the naturalist gained for his trouble.

Drury desired any insects that were rare, especially species that might prove new to European science. The avid entomologist planned to publish what he tentatively titled a "History of Insects" that would describe and illustrate each insect in his collection that had not previously been described by European naturalists. Drury believed that Britain's global commerce presented the nation's naturalists with unprecedented opportunities for acquiring nondescripts and thereby advancing natural knowledge. In the preface to the first volume of his *Illustrations of Natural History*, the entomologist declared that never before had naturalists had such opportunities to obtain new specimens "as in the present age." "All corners of the world are visited by our ships," he enthused. "The remotest shores of Europe, Asia, Africa, and America, are not unknown to our countrymen. . . . Every lover, therefore, of this study, must naturally hope, that such noble occasions of increasing the knowledge of nature, may not be neglected." The avid entomologist, for one, was not going to neglect the chance to add specimens from all corners of the world to his collection. If the globe's distant regions were known to his countrymen, Drury rationalized that there was no reason why their insects should not be as well. The naturalist recruited potential collectors from among British mariners bound for "the remotest shores" of Europe, Africa, Asia, and the Americas.[33] His ledger identified potential collectors among Britons bound between 1766 and 1773 for India, China, the Bay of Honduras, the West Indies, North America, West Africa, and West Central Africa.[34]

Although the scope of Drury's collection was global, in the years immediately following the discovery of the Goliath beetle, he particularly sought to acquire specimens from West Central Africa. The naturalist's desire for the Goliath beetle did not blind him to the many other interesting specimens likely to be found in Africa. He argued that European naturalists knew so little about African fauna that any African specimen gathered by one of his maritime collectors would have "intrinsic value" not only for him "but [for] Mankind."[35] Drury declared in 1768 to a correspondent in Sierra Leone that "I am more desirous of getting some [insects] from Africa than any other part of the world."[36] The frequency with which vessels bound for West and West

Central Africa appear in Drury's ledger attest to this desire. Travelers bound for West and West Central Africa account for one out of every three entries in his Aurelian ledger. While the slave trade and the plantation complex it enabled were crucial to the development of the Atlantic World, Britain's commerce with Africa accounted for a relatively small fraction of the empire's foreign trade. The frequency with which voyagers to West and West Central Africa appeared in Drury's ledger therefore reflected Drury's collecting priorities more than the geographic composition of Britain's global trade.[37]

Most of the voyagers to West and West Central Africa whom Drury identified as potential collectors were British mariners engaged in the slave trade. Drury's Aurelian ledger recorded fifty-five times that an individual bound for West or West Central Africa received boxes of collecting supplies from the naturalist. Thirty-eight recipients (or 69 percent) sailed to West or West Central Africa on board a slaving vessel. Three of the remaining recipients were naval officers assigned to the annual patrol of the West African coast to protect British trade and settlements. The business that brought the final fourteen individuals to West or West Central Africa is unknown. Very likely some of these were engaged in the direct trade in natural commodities rather than the slave trade.[38]

Regardless of the business that brought them to Africa, Drury's potential collectors typically began their journeys in London. Nearly nine out of ten identifiable voyages to West and West Central Africa in Drury's Aurelian ledger originated in London.[39] Mid-eighteenth-century London was a major port within the global trade to West and West Central Africa and was the second-busiest port within the British transatlantic slave trade. More important to Drury's collecting efforts, it was his home base. Proximity facilitated the recruitment of potential collectors and simplified the transport of the specimens they brought back. On the rare occasions when Drury's collectors sailed out of another port, the naturalist worried about what we might call today the "last mile problem." Drury fretted that delicate specimens could survive thousands of miles of transatlantic travel to reach Britain intact, only to be ruined by the jostling and vibration inherent to travel by horse-drawn carriage over imperfect roads. Drury therefore pleaded with correspondents in out-ports such as Liverpool to send specimens by sea rather than over land.[40]

Most of Drury's potential collectors sailed out of London because most of them were personally recruited by the London-based silversmith. Drury frequented the docks of London in search of collectors and collections. He befriended ship captains and surgeons, inviting them to his home upon their return to recount their voyages over a glass of wine. He met returning vessels to enquire whether any of the crew had brought back natural curiosities they were willing to sell. And, above all, he visited vessels before they sailed to persuade mariners to obtain specimens on his behalf. As Drury explained to a colonial correspondent, he visited "Ships in our River bound to distant parts" to distribute copies of "a few papers" to members of the crew. The "few papers" Drury distributed on board vessels in the Thames likely included his "Directions" and prints of specimens he desired.[41]

The most important aspect of "Directions" as a recruiting tool was its promise to pay sixpence for each specimen the naturalist received. Mariners took him up on this offer. For example, Drury recorded paying William Wallace's "Boy" one pound and one shilling for the specimens Wallace brought home on the slave ship *Africa* from Sierra Leone and the Gold Coast in 1768. Explicit promises of payment—and the fact that he kept these promises—helps to explain Drury's success recruiting mariners to become insect collectors. Prints, payments, and Drury's "Directions" thus became technologies for collecting collectors.[42]

The British mariners Drury collected were typically from the middling ranks within a merchant ship's hierarchy. The naturalist declared that he "always found . . . that the inferior Officers in the Ship were much properer persons" to serve as collectors "than either the common Sailor or the Captains; the latter have usually too much business of their own to attend . . . and the former have commonly too much of the Brute in them to comply with any design, even that beneficial to themselves."[43] Drury's ledger reflected his stated preference. Nearly two-thirds of the individuals Drury recruited to obtain insects in West and West Central Africa who can be identified fell into this group between common sailors and captains (table 5.1).[44] The thirty-seven individuals in Drury's ledger whose shipboard role can be identified included a boatswain, two carpenters, four ship surgeons, one second mate, six chief (or first) mates, and eleven individuals simply described as "mate."

A Mr. Drysale, ship carpenter on board the slave ship *Venus* in 1769, was among the slaving mariners who agreed to obtain specimens for Drury. Ship carpenters on any early modern sailing vessel were responsible for their vessels' structural soundness and were tasked with repairing and maintaining vessels' hulls, masts, and yards. On slave ships, they also built many of the technologies of control distinctive to vessels in the trade. Drysale would have constructed the *Venus*'s barricado, a tall, sturdy barrier that divided the ship's deck in half near its mainmast during the outward voyage to the African coast. The barricado served to separate male and female captives when

Table 5.1. Drury's collectors in West Africa and West Central Africa by occupation

OCCUPATION	NUMBER	(% of known occupation)
Boatswain	1	(2.6)
Captain	13	(34.2)
Carpenter	2	(5.3)
Mate	18	(47.4)
Second Mate	1	(2.6)
Chief Mate	6	(15.8)
"Mate"	11	(28.9)
Surgeon	4	(10.5)
Total of known occupation	38	(100)
Unknown	17	
Total	55	

Source: Drury Ledger, Letter-book of Dru Drury, folios 1v–3r, Natural History Museum, London.

on deck and provided a defensive barricade behind which the crew could retreat in case of insurrection. Outfitted with holes cut out for firearms and cannon, and featuring a small, closely guarded door, the barricado physically embodied the violence and coercive control that pervaded the slave trade. As historian Marcus Rediker observed, the barricado meant that "when the slaves were brought above, the main deck became a closely guarded prison yard." The outward voyage would have also been when Drysale built the bulkheads and platforms below deck that made possible a slave ship's inhuman crowding. Bulkheads divided men, women, and boys into separate sections of the space belowdecks, while the platforms vertically subdivided the already crowded space. Captive Africans would be "stowed" both above and below the platform, leaving each just over two feet of headroom.[45]

Drysale's voyage on the *Venus* as ship carpenter began on May 16, 1769. Two weeks earlier Drury had delivered a small box of collecting supplies for the carpenter to use on the slaving voyage to Anomabu on the Gold Coast and then to Jamaica. The captain and crew of the *Venus* "stowed" 230 captive Africans in the spaces Drysale subdivided belowdecks. During the weeks the *Venus* spent collecting human cargo along the Gold Coast, Drysale also collected specimens of butterflies, moths, beetles, crickets, and ants. Although Drysale's voyage on the *Venus* is his only entry in Drury's ledger, the provenance of specimens the carpenter collected and the dates when Drury received them suggest that Drysale may have obtained specimens on a second voyage, one to Sierra Leone. All told, the carpenter contributed specimens for twenty-six different species to Drury's museum, making him one of Drury's most prolific collectors in West and West Central Africa.[46]

The largest occupational group among the British mariners Drury recruited to collect in West and West Central Africa were ship's mates. These men best fit Drury's description of the "inferior officers" who were the naturalist's ideal mariner-collectors. Taken together, individuals described as second mates, chief mates, or simply "mate" accounted for nearly half of the identifiable individuals who Drury recruited to collect in West and West Central Africa. Further, two of the captains listed in table 5.1 began to obtain natural historical specimens while they were still first mates. Slave ships typically carried more mates than other sailing ships of similar size, meaning there were more mates for Drury to recruit. The presence of additional mates on slaving ships reflected both the trade's high mortality and the need for additional crew members to guard captive Africans on board. Mates on any deep-sea sailing ship were responsible for running watches and overseeing the daily operations of the vessel. The chief mate was second in command and would assume command in the event of the captain's death. Mates therefore needed to know how to navigate and how to keep the ship's books, both of which required some education. Drury argued that inferior officers "being persons who generally have received a midling education are more susceptible of Reason" and thus well suited as specimen collectors. Ship captains would have shared such an education, but Drury speculated that inferior officers were "more at leisure to perform a commission of this kind," especially "when accompanied with some advantage to them," such as the payment he offered for each specimen.[47]

Slave ship's mates who collected for Drury had less leisure than the naturalist imagined. In addition to the duties shared by mates on all sailing vessels, those on a slave ship were responsible for most of the day-to-day activities that made a slave ship "a strange and potent combination of war machine, mobile prison, and factory." Mates were tasked with "stowing" captive Africans belowdecks, checking the security of their shackles, overseeing their feeding and force-feeding, and bringing them above deck each day to exercise or "dance." They supervised the preparation of the captives' food, the distribution of medical care (without concern for captives' consent), and the prevention of suicides. In regions of West and West Central Africa where enslaved Africans were purchased from the ship's boats, the mate might share responsibility for this as well. Above all, mates were responsible for the security of the ship and the prevention of insurrections. Violence and terror were the primary means by which these objectives were achieved. The daily processes by which captive Africans were remade into commodities through violence and terror may have originated with the captain's orders, but they were implemented under the watchful eye of the mate.[48]

The British slave ship mates who oversaw the brutal treatment and commoditization of African captives were simultaneously Drury's ideal candidates to acquire butterflies and beetles in West and West Central Africa. Drury's collectors included a mate of the *Revenge*, bound for Senegal in January 1771 under the command of Captain Willcox. The *Revenge*'s mate made good use of the large box of collecting supplies Drury gave him; the mariner "bro[ught] home 10 Sorts" of insects later that year when the *Revenge* returned to London. Drury's investments in his potential collectors were often not this successful.[49] In 1767 he delivered a small box of collecting supplies to the chief mate of the *Good Intent*, commanded by James Gardner. The *Good Intent* sailed from Liverpool bound for Senegal but, as Drury recorded in his ledger, the ship and the naturalist's collecting supplies with it was "Lost—never heard more of it." Mr. Wright, mate of the *Duke of Marlborough* commanded by Captain Martin, sailed to "Guinea" in May 1767. Wright not only failed to collect but also lost the box and collecting supplies Drury gave him. Mr. Burn, chief mate of the *Hope*, proved similarly disappointing as a collector, but the naturalist recorded that he at least returned the box of collecting supplies, enabling the naturalist to reuse it by gifting to another traveler.[50]

Drury's most prolific mate-collectors gathered specimens over the course of multiple voyages. William Wallace, chief mate on the *Africa*, first agreed to obtain specimens on a slaving voyage to Sierra Leone and the Gold Coast in 1767. Drury delivered a small box of collecting supplies to the chief mate shortly before the *Africa* departed London's docks. The slaving vessel was commanded by Capt. John Stephens and carried a crew of forty at the voyage's outset. Under the supervision of Stephens and Wallace, the crew purchased 310 enslaved Africans in the Sierra Leone estuary and at Cape Coast Castle. During the four and a half months Wallace spent in Sierra Leone and the Gold Coast buying captive Africans, "stowing" them belowdecks, and employing violence and terror to prevent insurrection, he also gathered approximately forty-two specimens for Drury. Wallace sold his insect collection to Drury four days after the *Africa* returned to London. Six months later Wallace sailed again for the Gold Coast as first mate on the *Africa* under the command of Captain Stephens with a crew of forty. Stephens died during this second voyage, making Wallace the *Africa*'s new captain. Wallace commanded at least two additional slaving voyages, although these do not appear in Drury's ledger. Over the course of Wallace's slaving career, the mate-turned-captain continued to acquire African specimens for Drury.[51]

Slave ship captain Thomas Williams also began collecting specimens for Drury while a first mate. Drury delivered a small box of collecting supplies to Williams on June 15, 1767, shortly before Williams departed London as chief mate on the *King of Prussia*. The *King of Prussia* was bound for Bonny and then Puerto Rico under the command of Capt. Stretch Cowley. Like Wallace, Williams would have spent most of his time in West Africa overseeing the day-to-day operations of the vessel and the brutal processes by which captive Africans were treated as human commodities. He also gathered a small collection of insect specimens that Drury described as including some "very fine" specimens. Williams next sailed to West Africa in early 1769 as commander of the slave ship *Meredith* bound for Sierra Leone and then Grenada. Over the course of Williams's career he obtained specimens for at least seventeen different species of insects from coastal West Africa.[52]

Williams was part of a small cluster of collectors who began doing so while on board Cowley's ship. Cowley himself only collected two specimens, but mariners he commanded included more active collectors. William's replacement as Cowley's first mate—and Drury's collector—was a Mr. Reynolds. Drury delivered a large box of collecting supplies to Reynolds in September 1768 as he prepared to depart London on board the *King of Prussia*. The slaving vessel was bound for Bonny in the Bight of Biafra and the British Caribbean. Reynolds sold Drury thirty-seven African insect specimens when the *King of Prussia* completed its slaving voyage in September 1769. The following year, Captain Cowley again sailed to the Bight of Biafra, this time in command of a larger slave ship, the *Tartar*, with his son as first mate. The younger Cowley and the ship's surgeon, a Mr. Richards, promised to acquire specimens for Drury during the voyage. The *Tartar*, with its crew of fifty and two of Drury's large boxes on board, departed London in December 1769, bound for Bonny and Jamaica. It was a particularly deadly slaving voyage. Captain Cowley and his crew embarked 513 captive Africans in the Bight of Biafra, but less than half survived the voyage. Cowley and his son also perished on the voyage, leaving the second mate, William Thoburn, in command.[53]

The deadly voyage of the *Tartar* serves as a reminder of how Drury's collection of West and West Central African specimens was embedded in the violence and death that characterized the slave trade. Other reminders can be found throughout Drury's ledger. Drury recorded that in 1768 he gave a small box of collecting supplies to a slaving mariner whose "Ship was cutt off at Senegal." Drury indicated that the man survived the shipboard insurrection and returned to England "with a Large parcel" of specimens. On other occasions, Drury records the loss of a box of collecting supplies when the recipient, such as William Lightfoot of the slave ship *Providence*, "died on the Coast." The *Good Intent*, whose mate had promised to collect, was "Lost—never heard more of it." While to Drury these entries merely marked a missed opportunity to acquire specimens, to modern readers they should serve as reminders that collecting butterflies and beetles was at most a sideline to the violent and deadly business of collecting and commodifying captive Africans.[54]

CARGOES OF INSECTS

The *Hector*, commanded by Capt. Nonus Parke, completed its slaving voyage to Calabar and Dominica in September 1770. Two months later Drury learned of the *Hector*'s return to Liverpool. The naturalist was impatient to hear "the fate & success of [his] Box & Instruments for catching Insects." Had the slaving captain filled the box with specimens by following Drury's "Directions" and using his supplies? If so, "I shall be exceedingly pleased to find you have brought me a large Cargo." The naturalist "hope[d] nothing happened at Guinea to prevent my Box being filled." Drury's worries were for naught, as Captain Parke had indeed brought him a "large Cargo" of insects. On board the *Hector*—presumably packed in Drury's boxes—was a large collection of insects that had been gathered by Africans employed by Parke in Calabar. The collection included many species new to Drury, especially among its butterflies. Parke's entomological cargo was bittersweet for Drury, as many of its insects were badly damaged. The slaving captain blamed the "unhandiness" of the Africans he had employed as specimen-hunters. Drury, however, diagnosed the problem differently. Parke had not followed the injunction in Drury's "Directions" to wait until insects "are quite dead" before pinning them into a box for transport. The beetles in the box "with their claws" destroyed many of Parke's butterflies "by tearing them to pieces, leaving only their bodies & part of their wings which were next [to] the pins." To prevent such destruction in the future, Drury recommended that Parke ensure insects were dead before placing them in a specimen box. Further, the naturalist instructed that delicate species like butterflies should be segregated from beetles, or, at minimum, beetles should be pinned with their bellies facing up, thus limiting their reach and the potential damage they could do to their neighbors.[55]

It is unclear whether Parke implemented Drury's advice on his next slaving voyage. The slaving captain made one more voyage on the *Hector*, to Calabar and then Dominica in 1771. Shortly after the *Hector* returned to Liverpool, Parke informed Drury that he had a box of specimens for him but that poor health would delay his ability to send it. For more than six months, Drury impatiently waited for Parke's box to reach him in London. He wrote reminders and deputized another captain of his acquaintance to retrieve the box in Liverpool. Yet weeks later Drury still had not received it. "As I have not heard any thing more of the matter am afraid your bad health has occasioned their not being sent," the naturalist explained. He continued, "beg[ging] you would not think me troublesome if I desire you not to wait . . . but to send the Box by some other Vessel." If Drury did eventually receive the box, the specimens it contained must have been badly damaged or were duplicates of species already in the naturalist's collection.[56]

Over the course of Parke's career as a collecting slave trader, the ship captain acquired "perfect" specimens for twenty-one species of insects new to Drury's collection. Since Drury did not record damaged or duplicate specimens in his catalog, it is impossible to know how many more specimens Captain Parke acquired during the slaving voyages of the *Canterbury* and the *Hector*. All of Parke's specimens were gathered in West Africa, mainly in Calabar, and likely were obtained by local African collectors. As was often the case, the slaving captain's only acknowledgment of his reliance on African collectors came in the context of attempting to blame them for damaged specimens.[57]

Some of the African specimens Parke brought back on board the *Hector* were eventually depicted on the pages of Drury's *Illustrations of Natural History*. For example, plate 42 of the text's second volume (fig. 5.3) depicted three crickets: one from West Africa, one from the Bay of Honduras, and the third from an island off the eastern coast of Africa. Drury explained that he received the cricket labeled as figure 1 "from Callabar, on the Coast of Africa, situated in about 6 deg. North lat." The naturalist's manuscript catalog to his collection reveals that this specimen, *Gryllus caeruleus*, was obtained by Captain Parke during the *Hector*'s slaving voyage to Calabar in 1769. Drury's manuscript catalog collection similarly allows us to identify other specimens depicted on the pages of *Illustrations* that were originally obtained and transported by slaving mariners. For instance, the particular specimen of the *iphis* butterfly depicted in *Illustrations* Drury received from the slave ship captain

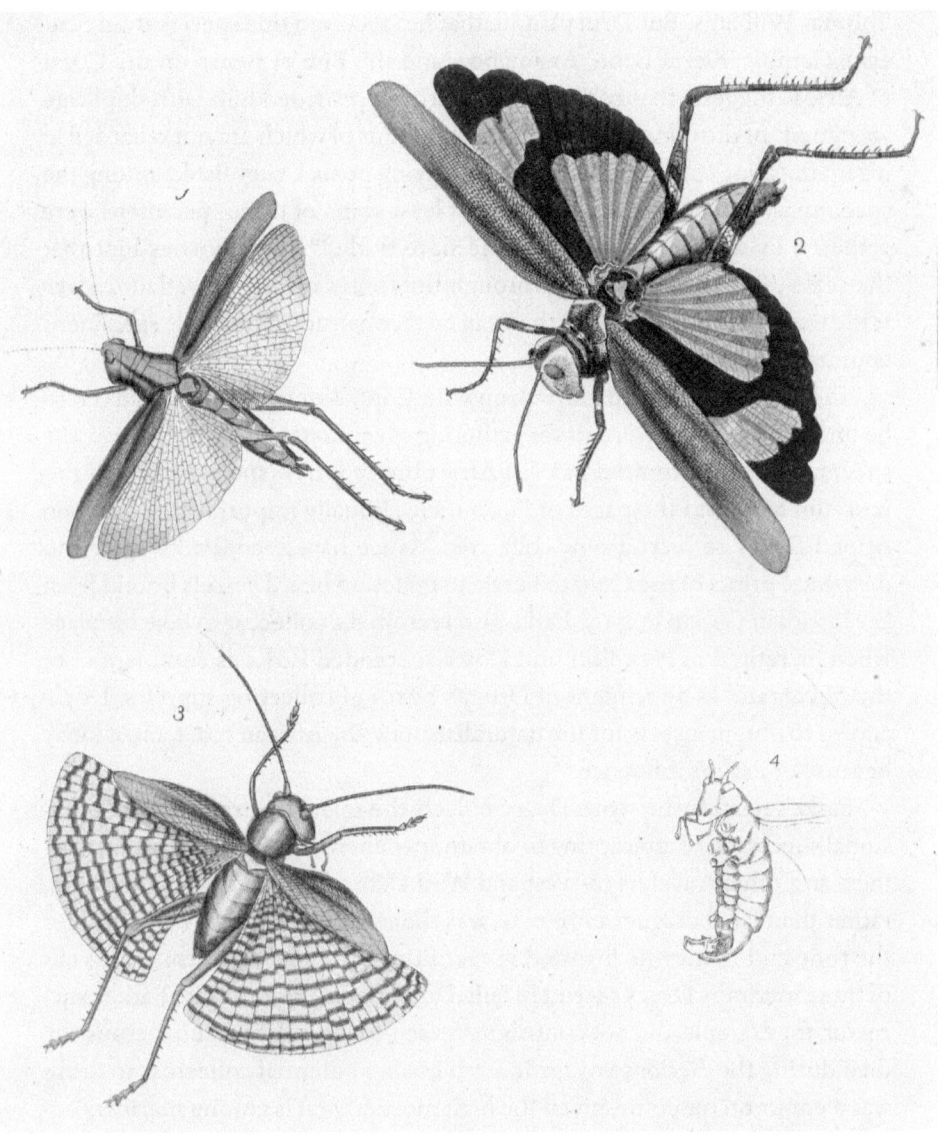

FIGURE 5.3. *Gryllus caeruleus*. Dru Drury's *Illustrations of Natural History* featured over forty insects collected by slaving mariners, including the cricket from Calabar labeled figure 1. Drury obtained the *Gryllus caeruleus* specimen from the slave ship captain Nonus Parke around 1769. Dru Drury, *Illustrations of Natural History*, London, 1773, Vol. 2, Tab. 42, RB 152307, Huntington Library, San Marino, California.

Thomas Williams. But Drury's note that he "received this species from Senegal, Gambia, Sierra Leon, Anamaboe, and the Bite of Benin on the Coast of Africa" suggests that other British mariners provided him with duplicate specimens of the *iphis* butterfly, the acquisitions of which are not recorded in his manuscript collection catalog. The slaving ports Drury listed among the specimens' provenance suggests that at least some of these specimens were gathered by mariners engaged in the slave trade.[58] These entries hint that the scale of Drury's collecting through the routes of the transatlantic slave trade was significantly greater than can be reconstructed from the specimens enumerated in surviving sources.

The relationship Drury struck up with Capt. Nonus Parke turned out to be one of his most important for gathering specimens from West Africa. The specimens Parke acquired in West Africa brought new species into Drury's museum and onto the pages of *Illustrations*. Equally important, the captain helped Drury to recruit new collectors. As we have seen, Parke agreed to distribute prints of the Goliath beetle to mates on board vessels bound from Liverpool to Gabon in 1769. Parke also recruited a collector to take his place when he retired in 1771. Edmund Doyle succeeded Parke as commander of the *Hector* and as a recipient of Drury's boxes of collecting supplies. Doyle agreed to obtain insects for the naturalist along the African coast, most likely because of Parke's influence.[59]

Parke's relationship with Drury reflects the many frustrations, and occasional successes, of attempting to obtain specimens by means of British mariners and other travelers to West and West Central Africa. Disappointment, rather than large cargoes of insects, was the usual return Drury received for the time and money he invested in recruiting potential collectors. Two out of three mariners Drury recruited failed to obtain a single insect. Parke's successor, for example, did not contribute insects to Drury's museum because he died during the *Hector*'s voyage in 1773. Losing potential collectors to death was a common outcome, given the high mortality rates among mariners engaged in the slave trade. During the months Drury awaited news of Parke's second voyage on the *Hector*, he began to worry whether the captain had survived the voyage to Calabar. The naturalist went so far as to write Parke's former landlady in Liverpool to ask whether the Captain Parke reported in the newspapers as having returned to Liverpool from Jamaica was Nonus Parke. "It being reported that Capt. Parkes died on the Coast is the reason of my troubling you with this," Drury explained.[60] More commonly, mariners failed to fulfill promises to collect for a host of reasons never recorded. "Brought nothing" thus became a common refrain throughout Drury's Aurelian ledger.

As Drury declared to Richard Cowley, mate of the slave ship *Tartar*, his efforts to acquire new specimens from West and West Central Africa were often met with "many dissappointments ... such as would have damped the hopes of any person less ardent in this pursuit than myself."[61]

Despite Drury's many disappointments, his reliance on the routes and personnel of the slave trade resulted in one of the best entomological collections of African specimens in eighteenth-century Britain. All told, Drury's museum included specimens for 209 "perfect" West and West Central African species of insects gathered by British mariners and other travelers.[62] Duplicates and damaged specimens would have brought this total much higher. To obtain these specimens, Drury recruited over fifty mariners and other travelers bound for West and West Central Africa in the 1760s and 1770s. His ledger documented that most of these travelers were mariners engaged in the transatlantic slave trade. Drury's offer to pay sixpence per insect was an attractive inducement for mariners and other travelers interested in finding additional ways to profit by their slaving voyages to West and West Central Africa.

Slaving mariners who collected insects for Drury seem to focus their collecting efforts exclusively on West and West Central Africa. Drury's printed "Directions" welcomed insects from any region of the world, and the routes of the British transatlantic slave trade would have brought Drury's collectors to American ports such as Jamaica and Dominica where enslaved Africans were disembarked and sold. Earlier in the century the slaving surgeons and captains who collected for James Petiver in West and West Central Africa also gathered specimens in American ports of disembarkation. Yet the British slaving mariners who collected on Drury's behalf appear to have only gathered specimens from West and West Central Africa; none of Drury's American or Caribbean species are credited to his mariner-collectors. It therefore seems likely that Drury instructed the slaving mariners he recruited as collectors to focus exclusively on African specimens.[63]

The discovery of the Goliath beetle in 1766 inspired Drury to focus his efforts on acquiring West and West Central African specimens. The pages of *Illustrations of Natural History* document the impact of Drury's redoubled efforts to acquire African specimens following the discovery of the Goliath beetle. The text's first volume (1770) largely depicted insects Drury acquired before the Goliath was found in the Gabon Estuary.[64] It included only four West or West Central African insects, representing just 2 percent of the species depicted in the volume. *Illustrations'* second volume (1773), by contrast, featured forty African specimens, comprising more than 20 percent of the text's total. The vast majority of the African specimens that Drury obtained from slaving mariners date to the period immediately following the arrival of the Goliath beetle in London.[65]

The cargoes of African insects Drury received from his mariner-collectors helped to make his entomological collection one of the most extensive in late eighteenth-century Britain. Even well into the nineteenth century, naturalists marveled at the diversity of specimens Drury assembled through the routes of British commerce and colonialism. The British naturalist John Obadiah Westwood defended his decision to produce a second edition of Drury's *Illustrations of Natural History* in 1837 based on the "extreme rarity of many of the insects figured therein." The second edition's new title, *Illustrations of Exotic Entomology*, highlighted what Westwood believed to be the text's two primary contributions: the high quality of its engravings and the inclusion of so many rare and foreign species on its pages. Westwood declared that "the acknowledged value of the figures contained in Drury's 'Illustrations' . . . continue up to the present day to be unique." Yet Westwood also apologized for the limitations of the descriptions that accompanied the engravings. They often lacked "the more minute characters, of which the present state of the science requires the investigation." Westwood provided additional details where he could, but his ability to do so was severely limited since the collection itself was no longer available to consult. By 1837 Drury's collection could only be seen in the pages of *Illustrations*. As so often was the case, the museum Drury spent a lifetime acquiring was dispersed in just a few days. Drury attempted to find a buyer willing to purchase the entire collection intact before he died. His failure to do so meant that a year after his death his heirs sold the 11,000 specimens in his collection by auction over the course of three days in May of 1805.[66]

CONCLUSION

In his quest to obtain a Goliath beetle, Drury distributed hand-colored prints to mariners on ships bound for West and West Central Africa, gave them boxes filled with collecting supplies, promised a sixpence for each insect he would buy from them, and assured new Aurelians that collecting insects would not interfere with the collection of captive Africans. Although these techniques failed to add a single *Goliathus goliatus* to Drury's museum, they did add scores of other rare African specimens to his collection. Drury's collection of insects by means of British mariners plying commercial and naval routes to West and West Central Africa required a fairly detailed understanding of the dynamics of British trading in the region. His letters reflected knowledge of West and West Central African geography, the rhythms of the British slave trade, and the ways collecting specimens might be integrated into the business of British naval and commercial ships along the African coast. The naturalist understood that as the British involvement with the transatlantic slave trade grew over the eighteenth century, so too did opportunities for naturalists to exploit the resulting commercial ties to gather the specimens that provided the basis of their scholarly pursuits.

The Goliath specimen that David Ogilvie acquired from the Gabon Estuary in 1766 remained the only known specimen of the *Goliathus goliatus* in Europe for the remainder of the century. Along with the rest of Dr. William Hunter's collection, it was bequeathed to the University of Glasgow after the Hunter's death. Today it is part of the Hunterian Museum's zoology collection. Although the pink markings have faded, the specimen otherwise remains in remarkably good condition. Drury never obtained a *Goliathus goliatus* for his own collection, but he did eventually acquire a different species of Goliath beetle. This specimen appeared as plate 40 in the third volume of *Illustrations of Natural History*. Drury noted that he did not believe it to be the same species as the Goliath depicted in volume 1 but that it was "undoubted non-descript." In the posthumous edition of *Illustrations* published in 1837, Westwood named this species *Goliathus druryi* in honor of Drury. Yet the rules of taxonomy give precedence to the earliest published name for a species, and another naturalist had described the same beetle a few years earlier. So today Drury's Goliath beetle is known as the *Goliathus regius*.[67]

Drury did not receive the *Goliathus regius* specimen from the fifty-five mariners he recruited to collect on his behalf along the African coast beginning in the 1760s. Instead, Drury's Goliath beetle was gathered by Henry Smeathman, a traveling naturalist sent to Sierra Leone to collect on behalf of Drury and a group of British patrons in the mid-1770s. As we will see in the next chapter, Smeathman spent nearly four years in Sierra Leone gathering specimens and studying the flora and fauna of the Upper Guinea coast. To do so, he relied on the local networks and infrastructure of the slave trade within Sierra Leone.

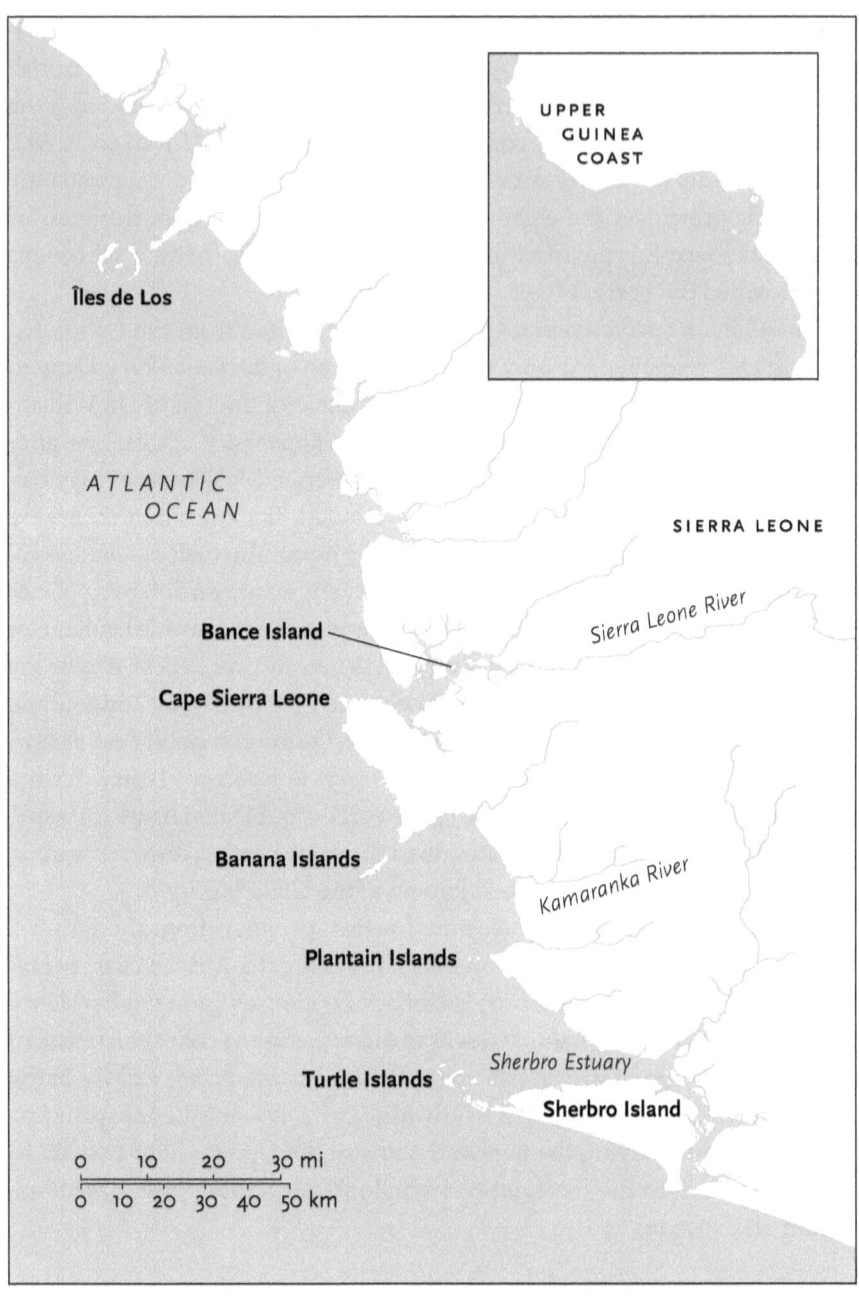

MAP 6.1. Locations of Henry Smeathman's collecting efforts in Sierra Leone, 1771–75. Smeathman's collecting efforts were largely confined to the areas of the Upper Guinea coast controlled by slave traders of his acquaintance: Sherbro, the Kamaranka River region, the Sierra Leone estuary area, and the Îles de Los.

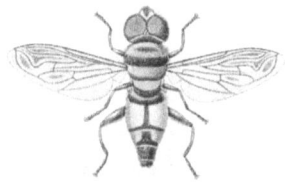

6.

A Flycatcher among Slave Traders

Early on the morning of December 13, 1771, Henry Smeathman stepped foot on the continent of Africa for the first time. Having already failed to make his fortune as a cabinet maker, insurance broker, wine and brandy merchant, and private tutor, Smeathman had left England six weeks earlier in the hopes that he might finally find fortune in Sierra Leone. The slave ship captains and traders whom Smeathman met there certainly understood how an Englishman might hope to exploit Africa to his own benefit. They were a bit more surprised that he planned to accomplish this through the collection of insects, plants, and other natural curiosities.[1]

Smeathman's first stop in Sierra Leone was the home of Capt. Robert Aird, a prominent slave trader to whom he had been "particularly recommended." Although the newcomer was eager to explore the area's flora and fauna, he first was obliged to present his credentials and "stand the fire" of the assembled slave ship captains. According to the self-described "flycatcher," the slaving captains viewed his butterfly-catching nets, forceps, and other collecting equipment with "an air of doubt & comtempt," and "could not conceive" why he came into the house "loaden with Grass & Weeds which grow round their very doors." Were the butterflies and other insects "for physic" (medicine), they inquired, "or to make pictures of?" They told Smeathman that while they had often seen pretty butterflies and "monstrous ... creeping things," "they never thought them *good for any thing*." "But what the Devil" did he plan to do with the grass? Did he not realize it would not keep green during the voyage back to England? What about the stones, clay, seeds, and nuts he had gathered? What did he propose to do with them? Surely he did not plan to eat them? "Who the duce would have thought such stuff worth the picking up?" "Well, Well," one captain concluded, "the longer one lives the more one learns. To think now of any body coming two or three thousand miles to catch butterflies & gather weeds. But every man to his calling."[2]

This description of the flycatcher's first day in Sierra Leone comes from the naturalist's own journal, essentially the first draft of his planned but never published travel account of the nearly four years he spent along the Upper Guinea coast.[3] In it Smeathman self-consciously presented himself as an enlightened traveler, poles apart from the base materialism and ignorance of the slave traders he encountered in West Africa. The naturalist's intended readers would have known very well why "such stuff" was "worth the picking up." The value of these specimens lay in their rarity, for relatively few West or West Central African species were known in European natural history. Smeathman's recounting of the captains' queries highlighted their unfamiliarity with the practices of natural history as well as their inability to view the world through any lens but that of potential profit. Smeathman concluded his narrative by declaring, "These & a hundred more stupid questions & impertinent remarks I was content to put up with & was not at all disatisfied to be taken for a fool by men of such profound heads."[4] With such comments, the naturalist sought to distance himself from the company he kept. Yet lurking beneath Smeathman's ironic and understandably contemptuous depiction of the slaving captains were similarities to his own situation greater than he would have cared to acknowledge.

The differences between Smeathman's collection of *naturalia* and slave traders' violent commodification of human beings were profound. While it would be a mistake to overlook these differences, it would also be a mistake to take Smeathman at his word that his efforts were divorced from the men that he met and the trade in which they were engaged. The naturalist tried to fashion himself as a gentleman motivated solely by the pursuit of natural knowledge rather than the desire for personal gain. However, Smeathman hoped to profit financially as well as intellectually from the years he spent in West Africa. Further, Smeathman's pursuit of natural history along the Upper Guinea coast was inextricably linked to local and transatlantic commercial networks, especially those of the slave trade. The naturalist spent approximately four years between 1771 and 1775 collecting and studying the flora and fauna of the Upper Guinea coast. Smeathman's journal and correspondence provide a unique window into the social and material practices of British natural history in West Africa at a moment when the British slave trade was at its height. They allow us to examine how a naturalist could exploit the local infrastructure of the slave trade to obtain, study, and transport specimens.[5]

The slave traders that the naturalist met at Captain Aird's house declared that they would be happy to have him as a guest on their vessels, where he would enjoy "fine sport" hunting the thousands of cockroaches he would find on board. In Smeathman's journal, these invitations were greeted with mockery and derision. The captains' inability to distinguish rare African insects from the common and well-known pests that infested their vessels served as yet another example of their ignorance of natural history. Yet the captains' willingness to accept Smeathman as a guest on their vessels proved essential to his search for rare African insects and other natural historical specimens. Slaving vessels—from large, transatlantic ships to small, indigenous canoes—became the naturalist's primary mode of transport within Sierra Leone. Most of the specimens he sent to England and the supplies he received in return traveled on the vessels of the transatlantic slave trade. More broadly, Smeathman's pursuit of natural history depended upon the social, political, and commercial infrastructure of the trade in human cargo and natural commodities. The slaving captains, traders, and agents Smeathman met in Sierra Leone formed his social circle and were his companions on collecting expeditions. When not collecting, Smeathman frequently dined, hunted, and even played golf with them. Perhaps partly as a result of his growing friendship with these men, Smeathman's attitudes toward slavery, the slave trade, and enslaved Africans proved more malleable than his early journal entries would have suggested.

"EVERY MAN TO HIS CALLING"

Smeathman reported that the slave traders he met at Captain Aird's house—"busily engaged in Solid pursuit of slaves, Ivory, Gum, Camwood, & other articles"—could not understand the errand that brought him there. The responses he ascribed to the slave ship captains suggested that this was not entirely true. As Smeathman's description highlighted, they were not exclusively engaged in the trade in human captives. The African trade, particularly in Sierra Leone, was never wholly synonymous with the slave trade.

Europeans also traded in natural commodities such as gold, ivory, and beeswax; medicines such as gum arabica; and dyes such as camwood.[6] Therefore, Smeathman's desire to commoditize the natural world of Sierra Leone for his own profit would have been somewhat familiar. While the slaving captains and traders might not have understood the finer points of specimen collection, they readily accepted the idea that there was more than one way to exploit Africa to one's profit. To paraphrase one of the slaving captains, they understood that the naturalist came to Sierra Leone to pursue his calling.[7]

For the first twenty-nine years of Smeathman's life, his calling was anything but clear. Like most Britons who sought their fortunes in West Africa, Smeathman left behind him limited economic prospects. The naturalist was born in Yorkshire in 1742 as the youngest of three sons. His father, a distiller and brandy merchant, died while Smeathman was still a boy. Smeathman left school early following his father's death and struggled to find an occupation that suited his aspirations, inclinations, and abilities. After a series of false starts, by 1771 he was employed as a tutor to a wealthy family in Birmingham. That summer Smeathman learned that the physician and naturalist John Fothergill wished to send "some lover of natural history" to collect specimens.[8]

Smeathman had been a lover of natural history from a young age. His sister-in-law recalled that even as a little boy, he would disappear for long stretches of time only to reemerge with collections of butterflies, seaweed, or cockle shells. As a young man Smeathman built a small collection of English insects including rare specimens such as the English purple emperor butterfly. His collection and avid interest in entomology became a coin of entry into natural history circles and the basis of his friendship with other collectors, including Dru Drury, whom we met in the previous chapter, and James Lee, a nurseryman and author of the popular guide to the Linnaean system, *Introduction to Botany* (1760).[9]

These friends and their connections facilitated Smeathman's new career as a traveling naturalist and specimen collector in Sierra Leone. It was from Lee that Smeathman first learned of Fothergill's desire to sponsor a collecting expedition. Smeathman asked the nurseryman to relay to Fothergill his eagerness to travel and "that a voyage to the coast of Africa would be exceedingly pleasing to me, as a country the least known to Europeans, and the most likely to afford a variety of new, curious, and valuable specimens in the three kingdoms of Nature." Left unsaid was his hope that such a voyage might also afford him fame as a naturalist and some measure of financial security. After a brief meeting between potential sponsor and prospective traveling naturalist, the two agreed that Smeathman would collect in West

Africa and that Fothergill would use his influence to recruit other sponsors for the expedition.[10]

Dru Drury did not need much convincing. Eager as always to add new insects to his museum, he quickly signed on as one of the expedition's patrons. Drury acted as Smeathman's primary agent in England, coordinating the disbursement of specimens, sending supplies, and alternately offering advice, chastisement, and sympathy in his steady stream of letters. Through the combined efforts of Drury, Lee, and Fothergill, the subscribers expanded to include William Pitcairn, a physician, and Marmaduke Tunstall, an English naturalist with a passion for ornithology. Joseph Banks, who just weeks before had returned from his own travels with Capt. James Cook on board the *Endeavour*, also agreed to support Smeathman's efforts. Margaret Cavendish Bentinck, Duchess of Portland, and the physician John Coakley Lettsom joined the ranks of Smeathman's sponsors in 1773.[11]

Each of Smeathman's sponsors initially contributed £100 toward the cost of his passage and collecting supplies. The traveling naturalist agreed to repay them in West African specimens. All the specimens Smeathman sent to Britain were to be appraised by three neutral experts and their value credited against his debt. After each sponsor received £100 in specimens, Smeathman was free to sell his subsequent collections on the open market for his "own advantage & benefit." Smeathman had in view the marketplace for natural historical specimens from the expedition's inception.[12]

Smeathman's subscribers functioned more like investors offering interest-free loans than traditional patrons of science. They expected that Smeathman would repay their investment through unusual and, hopefully, previously unknown African specimens. Similar to a commercial contract, their agreement with Smeathman was governed by explicit obligations and would be dissolved once those obligations had been fulfilled. In contrast, earlier eighteenth-century British collecting expeditions such as that undertaken by Mark Catesby were characterized by the subtle language, hierarchical relationships, and largely unspecified obligations of patronage. Although Catesby's patrons expected to receive specimens or drawings of American flora and fauna in return for their sponsorship, precisely what they would receive and in what quantities was left unspecified, leading to frustration and misunderstandings on all sides.[13] In the late eighteenth-century, the gift economy of patronage networks overlapped and occasionally conflicted with the commodity economy of the marketplace. Collectors could purchase rare plants, shells, insects, or other natural productions through a dealer, in a specialized London shop, or from the auction of another collector's museum.

Individuals like Smeathman who hoped to profit through the collection of natural historical specimens increasingly turned to the commodity economy of the marketplace to recompense their efforts.[14]

The growing marketplace for natural historical specimens encouraged collectors such as Smeathman and Drury to understand *naturalia* as commodities to be extracted. Smeathman and Drury shared a keen awareness of the potential markets for and value of West African specimens. In 1774, for example, Drury reported that he had seen two shells for sale at a London dealer's shop for forty pounds. He suggested that any new shells Smeathman discovered along the Upper Guinea coast might fetch similar prices.[15] Drury frequently reminded Smeathman to "never neglect an opportunity of sending something over [as] it will always fetch you money" since "at the worst" Drury could send them to Holland where he was certain they would garner high prices.[16] Precisely how high Smeathman's specimens were valued was an ongoing source of tension between the traveling naturalist and his sponsors. He complained, for example, that his first shipment of insects, which represented the labor of a year and a half, was only valued at eighty guineas.[17] Regardless of such disagreements, Smeathman and his subscribers shared an understanding of natural historical specimens as commodities with discernable monetary value.

The value of West African specimens was in part driven by how relatively few such specimens were available for study in Europe. As Smeathman observed to Fothergill when he proposed to travel to Sierra Leone, West Africa was among the places "least known to Europeans, and the most likely to afford a variety of new, curious, and valuable specimens." European mariners had regularly visited the region since the fifteenth century, yet European naturalists were familiar with only a small fraction of the area's natural productions. Travel accounts of the region occasionally included discussions of its flora and fauna, but the only extensive, systematic study of West African biota by a European naturalist was that by the Frenchman Michel Adanson. Adanson spent five years in West Africa, primarily in Senegal, and collected over 5,000 specimens of plants and animals between 1749 and 1754. After his return to France, Adanson drew upon his collections and observations to compose *Histoire naturelle du Senegal* (1757). Smeathman frequently cited the English edition of Adanson's text, which was published two years later in 1759.[18]

Smeathman's expedition came during a period of heightened British interest in West Africa. Parliament created the province of Senegambia just six years before the naturalist departed for Sierra Leone. The Crown colony of Senegambia encompassed the former British trading settlement on the Gambia River and the former French settlement of Senegal, which had been acquired as a result of Britain's victory in the Seven Years' War.[19] The English edition of Adanson's work was issued in this context. It was published the year after British forces captured French Senegal or, as the English preface to Adanson's text declared, "since the French settlements have been so gloriously reduced by the arms of Great-Britain."[20] The years when Smeathman visited Sierra Leone during the early 1770s were also marked by a dramatic increase in the slave trade along the Upper Guinea coast, largely dominated by merchants based in Liverpool. Smeathman and his patrons believed that there was much more to learn about West African natural history and that the need to do so was all the more pressing since British influence and commerce in West Africa seemed poised to expand.

Smeathman predicted that the West African specimens he collected would "make cent per cent" for his efforts. He calculated that by commoditizing the natural world of Sierra Leone and collecting along the routes of the local slave trade, he might finally find the competency that had thus far eluded him. Although ultimately he failed to do so, his plans reflected both the changing economy of British natural history and the growing belief among Britons that Africa could be exploited in new ways. In his journal and correspondence, Smeathman represented himself as a man of science free of the financial motivations that animated the slave traders with whom he socialized. Yet these sources also reveal that the naturalist shared the traders' desire to extract profit from his time on the Upper Guinea coast.[21]

AN INFRASTRUCTURE FOR COLLECTING AND ENSLAVING

Most West and West Central African specimens in British gardens, herbariums, and museums before Smeathman's expedition (see map 6.1) were acquired by mariners on slave ships or slaving agents stationed along the African coast. These slave traders integrated occasional collecting into the business that originally brought them to the African coast. Even Smeathman's French predecessor, Michel Adanson, was nominally employed by the Compagnie des Indes, although he was excused from his clerking duties. Smeathman, by contrast, arrived in Sierra Leone in 1771 as a traveling naturalist solely funded by wealthy and enthusiastic natural historical patrons. Smeathman's patrons included Quakers who were early supporters of abolitionism. Yet Smeathman discovered that his dependence on the local infrastructure of the slave economy was nearly as profound as that of the collecting slave traders who had proceeded him. Smeathman's pursuit of natural historical specimens along the Upper Guinea coast fundamentally relied upon the routes, people, spaces, and social customs of European trade with coastal Sierra Leone. The same social and commercial practices that facilitated Sierra Leone's trade in natural commodities and captive Africans could be employed to collect natural historical specimens. Further, Smeathman traveled within the Upper Guinea coast on slaving vessels, went specimen hunting with slave traders, used slaving forts as the bases from which to launch such trips, and often shipped the resulting collections back to Britain on transatlantic slave ships. Along the way Smeathman learned that the infrastructure of the local slave trade could frustrate as well as facilitate his collecting efforts.

The Banana Islands, where Smeathman made his home in West Africa, were controlled by John Cleveland, an Anglo-African slave trader and local ruler, and his younger brother James. As the naturalist explained to a correspondent, John Cleveland was "my friend & father here, as it is termed, because I was his visitor & stranger."[22] Smeathman followed local custom by putting himself under the protection of the elder Cleveland and offering his landlord gifts such as textiles, guns, or rum as both tribute and rent.[23] The landlord–stranger system offered travelers protection and integration into local society. In return, strangers offered their landlords gifts, an annual custom or rent, and taxes on any captive Africans they exported. Although over its long history the landlord–stranger system was extended to any traveler, by the late eighteenth century it was most closely associated with the European slave traders who relied upon it to facilitate their commerce in human beings. African elites, or landlords, provided European slave traders' housing, guaranteed their debts, and facilitated their trade. Often this included arranging temporary marriages between European slave traders and local African women. The reciprocal obligations between itinerant traders and the African elite who controlled the regions where they wished to temporarily reside were a key component of the social infrastructure of the slave trade in Sierra Leone.[24]

John and James Cleveland welcomed Smeathman to the Bananas "with the most friendly respect & civility." John Cleveland provided Smeathman and his English servant, David Hill, with a temporary home as well as permission to construct a more permanent dwelling.[25] The Clevelands introduced Smeathman to other members of the local African and Afro-European elite who controlled areas where the naturalist wished to collect. The younger Cleveland accompanied Smeathman on his first collecting expedition to the mainland. Cleveland and Smeathman traveled on a New England rum boat to the Kamaranka River to visit Charles Corker, one of the leading Anglo-African headmen and slave traders, and introduced Smeathman to the leaders of the towns where they stopped along the way. In ways large and small, Smeathman's landlord facilitated his integration into local society and promoted the objectives that brought him there, just as he would have done for a slave trader.[26]

The Clevelands likely helped Smeathman secure the labor of the free and enslaved Africans who were critical to the success of Smeathman's collecting efforts throughout his time along the Upper Guinea coast. African laborers accompanied Smeathman and James Cleveland to Coker Town and on subsequent collecting expeditions. Africans serving as hunters, guides, collectors, laborers, porters, and informants were crucial to Smeathman's work yet were often unacknowledged in his surviving papers. Smeathman also relied upon the labor of African collectors and hunters who traded with the naturalist for specimens they gathered on his behalf. For example, during a visit to a local slaving fort, Smeathman grumbled that "men, women & children crowd in" to his room "to stare, to ask questions and bring things to sell." He explained, "One says if I will give him powder and shot & lend a gun, he will get some fine birds, another will catch me insects, if I will give him some pins, another will fish for me [if] I will give him hooks."[27] The naturalist bartered for specimens with assorted trade goods sent by his sponsors including drinking horns, gold buckles, sauce pans, bullet molds, butchers' knives, pencils, axes, nails, scythes, camp ovens, and candlesticks. The variety of these items reflected the broad demand for manufactured goods on the Upper Guinea coast. A few items sent to Smeathman more directly reflected the role of the slave trade in Sierra Leone, such as the "6 doz. chains & Slave Chains" sent to the naturalist in May 1774.[28]

The Clevelands were most likely also involved in arranging Smeathman's marriage to the daughter of another member of the local Euro-African ruling elite, William Addoe. Smeathman's father-in-law controlled the region near River Sherbro, where the naturalist would spend many hours collecting insects and other specimens. Smeathman bragged that his "*Temporary* Father in Law" was "one of the greatest men in this country" without whose assistance he would have struggled to get settled. Smeathman's African wife would have played an even more important role helping the naturalist to integrate into local society. Interracial marriages between African women and European men were common along the West African coast. Such unions provided an important means by which Africans integrated Europeans into their societies and kin groups. As historian Pernille Ipsen has argued, they also provided

European men with physical and mental resources to survive in West Africa. European men such as Smeathman not only relied on their African wives to procure and prepare food for them and care for them when they were ill but also to serve as translators, cultural ambassadors, and trading partners. The social networks of Smeathman's wife, combined with those of his father-in-law and landlord, facilitated Smeathman's ability to travel within Sierra Leone on local vessels, to use trading forts as the bases from which to launch collecting trips, to barter for additional specimens, and to ship the resulting collections back to Britain on transatlantic slave ships.[29]

Such relationships did not develop by mere happenstance, nor did they begin with Smeathman's arrival in West Africa. Instead, they reflected the collective knowledge and commercial contacts of the experienced British merchants and mariners whom Smeathman and his subscribers consulted in the months preceding his departure from London. These men offered advice about where within West Africa he should travel, what supplies he would need, and what he might expect to encounter upon arrival. Their advice led Smeathman to focus his West African collecting efforts on Sierra Leone and to settle on the Banana Islands under the protection of the Clevelands. Perhaps more importantly, they provided the naturalist with introductions to the principal European and Euro-African slave traders in the region. For example, a Mr. Tiese drew up a list of the supplies Smeathman would need, recommended what presents to bring for local rulers, offered to arrange his housing, and procured letters of introduction to prominent traders in Sierra Leone. In his letters and journals, Smeathman often noted that he had been "particularly recommended" to the various slave traders he met. The extensive advice and assistance Smeathman received illustrates that while natural historians might have considered the region little known, British slaving merchants and mariners had known it well for decades.[30]

British slaving merchants would have known that European trade with West Africa varied extensively in terms of its local focus and structure. The structure of the slave trade along the Upper Guinea coast was deeply influenced by its physical environment. Numerous mangrove-lined rivers, marshy low plains, and powerful tides dominated the region. It was notorious for its bottom-ripping reefs, extensive shoals, and sandbars that blocked the entrance to rivers. The combination of these environmental factors meant that most travel along Sierra Leone's coasts and estuaries was in small boats. Oceangoing slaving vessels typically anchored off the coast, while the ships' longboats traveled between transatlantic vessels and slaving settlements along the shore. Traders typically brought captive Africans from more distant parts of the region directly to transatlantic vessels in a practice known as the ship trade. Some slaving captains opted for efficiency over price and called at the region's bulking centers in lieu of engaging in the ship trade. Bulking or warehousing centers were islands just off the Upper Guinea coast where local traders and transatlantic captains could find a range of services as well as a concentration of manufactured goods and captive Africans for sale. Slaving captains who opted to trade at bulking centers calculated that the premium charged for enslaved Africans and supplies at bulking centers would be offset by a shorter stay on the African coast, with its always present threat of sickness and death for crew and captives alike.[31]

Both the ship trade and bulking centers required extensive local maritime networks linking smaller trading posts with each other, with larger slaving factories, and with oceangoing vessels. European, Euro-African, and African slave traders along the Upper Guinea coast and its estuaries frequently exchanged captives, African commodities, and European manufactured goods with one another. All of this commercial activity required a steady circulation of ships' longboats, indigenous canoes, and other small vessels. Although Smeathman at times hitched a ride on transatlantic vessels, the naturalist more frequently traveled on smaller coasting vessels and indigenous canoes whose constant movements throughout the region were central to the slave economy.

Larger factories that functioned as bulking centers were major hubs within the local economy and attracted a steady stream of vessels large and small. They also tended to concentrate Europeans visiting and living in the region. Bulking centers therefore served as bases from which Smeathman could launch collecting expeditions as well as places where he might find transport and companionship for these trips. A collecting expedition Smeathman undertook during the summer of 1773 with a revolving cast of slaving agents and captains as his companions illustrates the ways in which he relied on Sierra Leone's bulking centers and extensive maritime networks. The naturalist spent at least five weeks of the nearly three-month trip at the region's two major bulking centers: Bance Island and Îles de Los.

Smeathman's first stop on his 1773 trip was Bance Island, located fifteen miles upstream from the mouth of the Sierra Leone River. Bance Island was operated by six London merchants who did business under the name Grant, Oswald & Co. It functioned as a "general rendezvous," where slaving vessels flying the flags of many European nations could be found anchored. European vessels would stop to purchase enslaved Africans, load ship stores, undertake repairs, and inquire about local trading conditions.[32] When Smeathman visited in 1773, the island would have been a hive of activity engaged in various aspects of the transatlantic slave trade. The years between the Seven Years' War and the American Revolutionary War were profitable ones for Grant, Oswald & Co. During this time the company employed approximately 35 Europeans in addition to the 142 enslaved Africans who permanently labored at the factory. These individuals worked as blacksmiths, coopers, surgeons, armorers, sailmakers, storekeepers, guards, and a host of other occupations. When captive Africans, visitors, and temporary residents are added to the island's free and enslaved permanent labor force, the population of Bance Island could at times exceed 700 souls. Ships anchored in its harbor ranged from transatlantic vessels to smaller sloops and canoes used to travel on local rivers and creeks. These smaller vessels connected Bance Island to its many out-factories, where the company purchased most of the captive Africans it sold.[33]

During the fortnight Smeathman spent on the island, the two agents who oversaw the fort's operations were focused on the imminent departure of a vessel owned by one of Bance's principal investors. The naturalist noted regretfully that the agents were too "busy dispatching the Brig Amelia" to undertake any collecting excursions. The *Amelia* was bound for Charleston, South Carolina, where only 287 of the 352 enslaved Africans who began the terrifying journey across the Atlantic disembarked.[34] In preparation for the vessel's transatlantic crossing, the agents would have overseen the embarkation of African captives and loading of ship stores, provisions, and other supplies. Smeathman, however, made no mention of this flurry of activity. Instead, he described the social life at the trading factory. He reported that he and the agents and assembled ship captains amused themselves "in the cool of the afternoon" playing whist, backgammon, or even golf on a course specially constructed on the island. They spent their evenings "with the bottle & pipe very chearfully." Smeathman's mornings must have been spent in part writing letters. The naturalist took advantage of the frequency with which transatlantic vessels called at the island to post a series of letters describing Sierra Leone's flora, fauna, and social customs to European patrons, naturalists, and other correspondents.[35]

After two weeks socializing and letter-writing at the fort, Smeathman and half a dozen companions departed Bance Island to hunt, collect specimens, and visit other Europeans along the Upper Guinea coast.[36] To do so, they relied on a series of different commercial vessels. The group initially traveled on a small coastal vessel that accompanied the *Amelia* out of the mouth of the Sierra Leone River and then headed south to Cape Sierra Leone. Smeathman's party next sailed to Whiteman's Bay, where they dined on board Captain Tittle's slaving vessel. In the morning they went shooting and collecting on Cape Sierra Leone near modern Freetown, most likely using smaller local vessels. This time Smeathman reported he had "tolerable success" obtaining specimens. After another dinner on board Tittle's vessel and "a chearful afternoon" with the assembled party, Smeathman and his assistants sailed on Captain Bishop's schooner for the Îles de Los. Arriving there two days later, Smeathman, Berlin, and Hill transferred to the slave ship *Africa*, commanded by Captain Wilding, an agent employed by the Liverpool slaving merchant Miles Barber.[37]

Slaving vessels provided housing as well as transport for Smeathman during his collecting expeditions. The slave ship *Africa* served as Smeathman's living quarters during the three weeks he and his companions spent searching for specimens on the Îles de Los in the summer of 1773. His description of the sleeping arrangements on board the vessel and more broadly of the slave ship itself are among the only handful of explicit references he made to the slave trade upon which he so frequently relied. "Alas! what a scene of misery & distress is a full slaved ship in the rains," he wrote. "The clanking of chains, the groans of the sick & the stench of the whole is scarce supportable." Smeathman explained that frequent wind, rain, and thunder prevented him and his assistants from spending much time on deck. Instead, they "sweat & stew in a cabbin two thirds filled with cloth of various kinds." The *Africa*'s cabin, with its trade goods and visitors, became even more crowded at night when the captain and some of the vessel's female captives and children added to its denizens. Smeathman, Captain Wilding, and the captain's other visitors occupied the cabin's entire floor, while female captives "lay upon the lockers or any where that will admit half their length" and children slept on chairs or any other space they could find.[38] Disease spread quickly on the crowded vessel. Captives, crew, and visitors suffered from a combination of "an epedemic kind of Measles," worms, the flux, and "a very terrible fever." According to Smeathman, almost two enslaved Africans a day died from the measles alone. The Europeans on board were not immune from the epidemic. Captain Wilding and Berlin were among those taken ill. Smeathman confided to his European correspondents his fears that the botanist's constant drinking may have ruined his constitution. One week later, Berlin was dead, having been on the Upper Guinea coast for just two months.[39]

After burying Berlin in the Îles de Los, Smeathman and Hill continued on board the *Africa* as it made its way south toward the Banana Islands. During the nearly month-long return trip, they continued to collect specimens and observe the region's natural productions from the decks of the slave ship. The insects and half dozen bulbs Smeathman and Hill gathered during this time suggest that they also made occasional trips to the shore. Smeathman remarked on the diversity of grasses and edible plants found near abandoned settlements and speculated on the commercial possibilities of a land so lush. By mid-July, Smeathman and Hill were back home on the Banana Islands.[40] Smeathman spent nearly three months traveling, hunting, and socializing with slaving captains, factors, and agents in Sierra Leone. Most, he reported, were renegades, gamblers, drinkers, and uncivilized boors who stirred their tea with dirty knives. Yet they also provided the assistance, transport, and hospitality that were crucial to his collecting efforts. The naturalist's nearly three-month expedition in the summer of 1773 illustrates the myriad ways in which he relied upon the personnel and infrastructure of the slave trade along the Upper Guinea coast.[41]

Smeathman had little choice but to depend upon the maritime infrastructure of local commerce because the naturalist had no boat of his own. Coastal Sierra Leone's riverain environment meant that most travel was by boat. Smeathman told the British naturalist Thomas Pennant in May 1773, "There is no travelling but by water, a circumstance I was not acquainted with." He blamed the centrality of maritime travel for why it took him eighteen months to visit Bance Island for the first time, despite it being "within thirty leagues of my habitation." The want of a boat of his own became the naturalist's frequent complaint as well as his reoccurring excuse for why his collection was not larger. For example, he reported that "the great Oxen, & indeed an inconceivable quantity of wild beasts, by all accounts, abound on the mountains of Sierra Leone," but he had not been able to make an excursion there to investigate these accounts without access to a boat. Traveling on local commercial vessels confined Smeathman's collecting to the vicinity of trading posts; areas like the mountains of Sierra Leone remained inaccessible to him if none of his acquaintances had a reason to visit nearby. The naturalist's collecting was largely confined to the areas of the Upper Guinea coast controlled by slave traders of his acquaintance: Sherbro, Smeathman's father-in-law's territory; the Kamaranka River region, controlled by the Euro-African trader Corker; the Sierra Leone estuary area; and the Îles de Los.[42]

The local and transatlantic mariners on whom Smeathman depended were also a frequent source of frustration for the naturalist. He blamed them for the many failures and mishaps he experienced along the way. At heart, Smeathman's complaints were one of a naturalist not in control of his collecting expeditions. Smeathman fumed that mariners "think you are much oblige[d] to them for your passage and will not give themselves the least trouble about any thing belonging to you so that sometimes you can scarce get them to lift your baggage out of the boat." European, Euro-African, and African mariners often did not treat the naturalist's specimens with the care he thought they required. Specimens of plants were tread upon, boxes of insects kicked about, shells smashed to pieces, and "everything injured." Smeathman accused some slaving mariners of deliberately sabotaging his collection by trying to slip specimens overboard when they thought he was not looking. Despite the advice he received in London from experienced slave traders, Smeathman was not fully prepared for the many challenges of traversing the maritime landscape of the Upper Guinea coast.[43]

Transporting specimens back to England on oceangoing vessels presented a separate set of challenges for the naturalist. Smeathman's sponsors urged him to ship his specimens on vessels that would return directly back to Europe from Africa. They worried about the damage sustained by specimens during the much longer voyage by way of the Caribbean. In November 1772, for example, Drury fretted about the condition of a shipment sent via the transatlantic slave trade. He predicted that as he penned his letter, Smeathman's shipment of specimens was "sweating in its voyage round by the West Indies."[44] Yet Smeathman reported that he encountered few vessels engaged in the direct trade and fewer still that would take his specimens as cargo. Instead, he relied on the transatlantic slave trade that dominated British commerce with West Africa in the late eighteenth century. The fate of Smeathman's second shipment of insects reinforced Drury's concerns about the transatlantic route. Smeathman placed his collection on board a slaver bound for Dominica under the care of Capt. Francis Barre, an acquaintance of Drury's. Upon arriving at the island, the vessel was condemned and its cargo divided between other vessels headed to England. Smeathman's box made its way back to London, but it sat at a merchant's warehouse for over a month. The shipment was simply labeled with Drury's name. Without Captain Barre present to provide more information, no one knew to which Drury the box belonged.[45]

Drury's fellow patron John Fothergill was perhaps even more committed to the idea that Smeathman should only use the direct trade between England and Africa. Fothergill's strong abolitionist beliefs made him wary of relying upon the slave trade. Vessels engaged in the direct trade typically trafficked in natural commodities rather than in human beings. In 1774 Fothergill blamed the relatively few specimens that subscribers had received from Smeathman on the difficulties of transportation by means of the slave trade. He explained to Carolus Linnaeus that the vessels on which Smeathman sent his specimens, "Laden with the wickedest of cargoes—men torn from everything that makes life worth while—they proceed to the West Indian Islands, and after a long delay return thence to England. Because of this long detour, everything dies." The Quaker physician proposed to recall Smeathman and to send him out a second time, this time in command of his own vessel that could directly return to England after collecting "an immense treasure of natural objects" in West Africa. Fothergill believed this would enable Smeathman to engage in natural history without relying upon the routes, vessels, and personnel of the slave trade.[46]

Further complicating Smeathman's reliance on transatlantic slaving vessels was the difficulty finding passage on board ships to London. During this period, Liverpool had the dubious honor of being the most active slave-trading port in Britain. More than seven times more enslaved Africans were transported to the New World on ships beginning their journey in Liverpool than in London in the 1770s.[47] The dominance of Liverpool was even more pronounced among British slave traders on the Upper Guinea coast. Yet all of Smeathman's sponsors were based in London. They had nearly as much trouble finding London ships willing to take supplies to Smeathman as Smeathman did finding direct trading vessels in Sierra Leone. Smeathman's family connections provided the solution. The naturalist's uncle, John Smeathman, was the collector of excise for the port of Liverpool during the years that the naturalist was in West Africa. Ship captains were happy to curry favor with the port official by loading a chest or two for his nephew on their vessels. According to one of Smeathman's patrons, his uncle's position was a "most Lucky circumstance" because it enabled Smeathman's subscribers "to send [his] Goods with great Ease" on Liverpool vessels, while the London ships were "very unwilling to take any thing."[48]

Smeathman continued to rely on both the local and the transatlantic slave trade throughout his four years in Sierra Leone. He ignored Drury's and Fothergill's pleas to return to England, recover his health, and prepare for a second expedition that would allow him to be independent of the "2 legged Brutes" engaged in the slave trade along the Upper Guinea coast.[49] Instead, Smeathman argued that he needed greater, not less, involvement with the slave trade in order to achieve his natural historical objectives. Smeathman agreed to become an agent for a leading Liverpool slave merchant during his last year in Sierra Leone. The naturalist declared that his position working for the Liverpool merchant William James would "be the completion of my fortune." Smeathman argued that his "collecting etc. instead of being hindred, will thereby be promoted." He noted that his new employer had thirty-seven vessels engaged in the African trade. As these ships "will be constantly calling upon me homeward bound, opportunity of sending my collections will be frequent."[50] It is unclear whether Smeathman availed himself of the opportunity to ship his collections on board James's vessels. What is clear is that by accepting the position, the naturalist had brought his relationship with local slave traders full circle. Whereas once he had taken great pains to distance himself from slave traders, he now eagerly joined their ranks and declared that doing so would enhance his pursuit of natural history.

THE COMPLETION OF A (NATURAL HISTORICAL) FORTUNE

Smeathman explained his decision to become an agent for the slaving merchant James both in terms of completing "his fortune" and enhancing his pursuit of natural history. His natural historical collecting along the routes of the slave trade in Sierra Leone yielded specimens for his patrons and produced the material basis for his own natural historical investigations. Pressed plants, pinned beetles, stuffed birds, and delicate shells were all means to repay his debts to his sponsors and, potentially, to profit from the marketplace for rare specimens in Europe. Smeathman also hoped that his collecting efforts would establish him as an accomplished naturalist in his own right. The specimens he gathered and the observations he made along the Upper Guinea coast enabled the natural historical studies he undertook both in the field while in West Africa and in the cabinet after his return to Britain.

Smeathman's most extensive investigations were those into the West African termites whose large nests dotted the landscape. Like his counterparts in Europe, Smeathman's pursuit of natural history in Sierra Leone represented an active program of collecting, observing, and experimenting on the organisms that he encountered. The naturalist sought to understand the habitats, anatomy, and behaviors of the species he observed.[51] Smeathman used chemical analyses to try to understand the composition of termitaries. He calculated the relative size and weight of a mature queen compared to a common laborer "by carefully weighing and computing the different states."[52] Microscopic analysis helped him to more fully understand the function of the three different compartments within termitaries: the royal chamber, the nurseries, and the magazines containing the colony's food. Smeathman observed that the nurseries were "always slightly overgrown with mould" and contained "small white globules about the size of a small pin's head." Upon examining these white objects under his microscope, he discovered that they were not eggs, like he originally assumed, but rather a species of mushroom cultivated

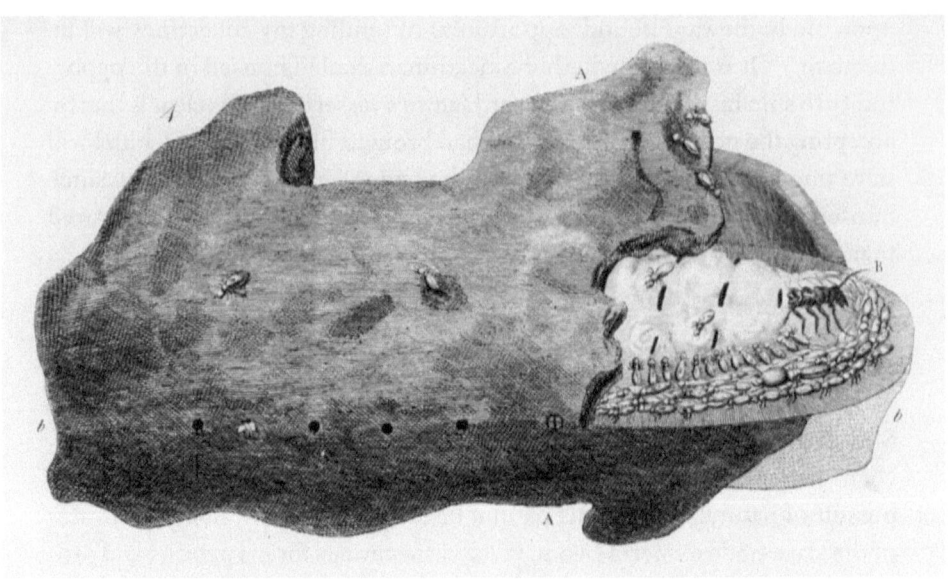

FIGURE 6.1. Royal chamber of a termite nest. Henry Smeathman's "Some Account of the Termites" (1781) depicted the royal chamber of a termite nest from multiple angles as well as its appearance upon first being opened by the naturalist. The angle pictured here highlights the enormous size of the *Termes bellicosus* queen, compared to her "attendants running round her." Henry Smeathman, "Some Account of the Termites, Which are Found in Africa and Other Hot Climates. In a Letter from Mr. Henry Smeathman, of Clement's Inn, to Sir Joseph Banks, Bart. P. R. S.," *Philosophical Transactions* 71 (1781), Tab. 8, RB 98681, Huntington Library, San Marino, California.

by the insects.[53] To get a better look at royal chambers, Smeathman removed them from their nests and placed them in a glass bowl in his home, allowing him to observe the colony's king, queen, and their "attendants" at his leisure (fig. 6.1). Smeathman's investigation of Sierra Leonean termites was built upon an active program of observing, manipulating, and experimenting.[54]

Smeathman's observations and investigations resulted in his "Some Account of the Termites, which are found in Africa and Other Hot Climates," first published in 1781. The text continues to command respect among entomologists for offering a detailed explanation of termite behavior that has been borne out by subsequent research.[55] Smeathman's "Some Account of the Termites" sought, above all, to reveal the hidden world of termites.[56] While the text opened with brief taxonomic descriptions for the five different species Smeathman collected, systematics functioned as merely prelude. As historian Mary Terrall argued, in the eighteenth century the "natural history of animals, especially insects, was not primarily a taxonomic endeavor." Naturalists such as Smeathman strove to understand the intricacies of insects' behavior, habitat, ecological role, and anatomy at a level of detail well beyond what was necessary for classification.[57] "Some Account of the Termites" provided the first detailed look into termites' social organization and life cycle as well as the internal architecture of their nests (see fig. 6.2).

The towering termitaries had not escaped the notice of previous travelers to the Upper Guinea coast. However, as Smeathman explained, "no one but myself has probably had time and opportunities enough to make" an extensive study of the insects.[58] Only by opening dozens of termitaries over the course of years could Smeathman piece together the structure of the termite nests he encountered along the Upper Guinea coast. The ubiquity of the termite nests on the Banana Islands and the adjoining mainland increased Smeathman's opportunities, with Cleveland's tacit approval, to study the nests' structure. The naturalist explained that "it was scarce possible" to stand in a clearing on the Banana Islands or the adjoining mainland "where one of these buildings is not [to] be seen within fifty paces."[59] Smeathman wanted to understand not only what the interior of a termitarium looked like but also how it came to look that way. Of what was it built? How was it maintained and repaired? What would happen if a hole was cut into it? How long would it take its resident termites to repair it? With so many termitaries nearby, Smeathman frequently repeated his incursions in the hopes of answering these questions. He reported, for example, that termites always traveled underground or within covered pathways because they otherwise made easy prey for ants and other predators. The naturalist observed that if he destroyed a few inches of the pathway, "it is wonderful how soon they rebuild it." A section of pathway three or four feet in length would be repaired overnight, but if the same section was destroyed several times, "they will at length seem to give up the point."[60] Smeathman's termitary investigations became part of his regular routine as he visited the same sites over many months to assess the progress the termites had made in rebuilding their nests.

Determining termitaries' interior architecture proved no easy task. As Smeathman and his African assistants dug into the termites' nests, the moist but brittle clay often crumbled. Further, the nature of termitaries' construction created "a kind of geometrical dependance" between individual chambers so that "the breaking of one arch pulls down two or three." The activities of the termites themselves also prevented a clear view. Smeathman described termite soldiers streaming out of the attacked termitarium "who fight to the very last, disputing every inch of ground" leaving those attempting to

FIGURE 6.2. Interior of a termitarium. The iconic engraving of the interior of a West African termitarium from Henry Smeathman's "Some Account of the Termites" (1781). The African man in the foreground pointing at the termite nest highlights the critical role played by African labor and knowledge within Smeathman's pursuit of natural history in Sierra Leone. Henry Smeathman, "Some Account of the Termites, Which are Found in Africa and Other Hot Climates. In a Letter from Mr. Henry Smeathman, of Clement's Inn, to Sir Joseph Banks, Bart. P. R. S.," *Philosophical Transactions* 71 (1781), Tab. 7, RB 98681, Huntington Library, San Marino, California.

break open the nests with bloody feet and legs. Once the termite soldiers deemed it safe, the laborers would immediately begin to repair the damage, in the process further obscuring Smeathman's view. The iconic image of a cross-section of a termitary included in "Some Account of the Termites" provided the reader with an unobstructed view unlike any that Smeathman ever experienced. In reality, "the whole internal fabric" of the termitary could only be seen, Smeathman explained, "by piece-meal."[61]

In the center of the iconic image of a termitarium's interior stands an African man in a relaxed posture, holding a hoe in one hand and pointing to the opened termitary with the other. The man appears to have just dug into the termitary (see fig. 6.2). Although the accompanying text tells us nothing about the man, his presence signals Smeathman's reliance upon free and enslaved Africans throughout his stay in Sierra Leone. The image highlights the physical labor undertaken by Africans in the process of opening termitaries.

Local men and women assisted Smeathman's pursuit of natural history along the Upper Guinea coast in a myriad of other ways. They likely undertook many of the same duties fulfilled by Black and Indigenous individuals whose labor, skill, and knowledge were crucial to the work of naturalists throughout the Atlantic World.[62]

African entomological knowledge was responsible for some of the most important contributions to the scientific literature made by "Some Account of the Termites." Termite reproductives (queens and kings) can be thirty times the bulk of a laborer from the same colony. With large, conspicuous eyes and four wings that only develop at the beginning of the rainy season, reproductives share little resemblance with laborers and soldiers of the same species (see fig. 6.1). Linnaeus and other naturalists therefore assumed that the termite only contained the two smaller, wingless castes. Before the publication of "Some Account of the Termites," termites were classified as *Aptera*, insects without wings. Smeathman declared that the reproductive "differs so much from its form and appearance in the other two states, that it has never been supposed to be the same animal, but by those who have seen it in the same nest; and some of these have distrusted the evidence of their senses." Smeathman learned that the termite had a winged stage in its life cycle from local Africans, who told him that all three castes "belonged to the same family."[63] Yet "it was so long before I met with them in the nests myself, that I doubted the information which was given me by the natives." To corroborate the information gleaned from his African informants, Smeathman offered his own eyewitness testimony as well as the drawings and specimens he brought back to Britain with him. Smeathman represented himself as the ultimate author of new matters of fact about Sierra Leone's natural world. Smeathman suggested that it required his experimentation and observations to turn African knowers' expertise into credible natural knowledge.[64]

The four years Smeathman spent pursuing natural history in Sierra Leone yielded a significant collection of natural and man-made curiosities. Before departing the Banana Islands in 1775, Smeathman asked Drury to procure £500 in marine insurance on his behalf. His detailed request provides a sense of the collection's extent. The collection included "great many dried plants, reptiles, &c. in spirits," fossils, shells, insects, stuffed birds, pieces of termite nests, and a small quantity of what he called "some very curious" fabrics, African utensils and other man-made items.[65] From the moment of their arrival in Sierra Leone, Smeathman, Hill, and Berlin had marveled at the diversity and fecundity of the Upper Guinea coast's natural world. Berlin boasted that he discovered three new species less than fifteen minutes into his first collecting excursion. Berlin's death just months after his arrival in West Africa frustrated Smeathman's ambitions for his botanical collection. Yet the collection was still rich enough for Smeathman to present Joseph Banks with nearly 600 unique botanical specimens, most of which were African nondescripts. Smeathman claimed that many of these represented new genera, not just new species. Appropriately for a naturalist whose primary interest was entomology, the collection was even more extensive when it came to insects. Smeathman boasted that his entomological specimens "have enriched most of the cabinets of Europe with singular and beautiful genera and species." Drury's cabinet was among those that most benefited from Smeathman's efforts. Drury credited Smeathman for 710 of the insect species represented in his museum, a count that does not include duplicate specimens. Specimens gathered by Smeathman included many that were described in the third volume of Drury's *Illustrations of Natural History* (1782). As Smeathman's boast to have enriched the cabinets of Europe implied, the museums that benefited from his expedition to Sierra Leone were not just those belonging to his patrons. The London physician William Hunter acquired at least 75 insects that can be traced to Smeathman's expedition. His brother, John Hunter, dissected termites originally collected by Smeathman and added these to his anatomical collections.[66]

The specimens that reached Britain intact, however, were only a fraction of those Smeathman and his assistants acquired. Contrary to his patrons' expectations, he sent relatively few specimens to Britain while he remained in West Africa. The naturalist instead planned to accompany them so that he could personally ensure their safety. However, Smeathman fell ill during the voyage to the Caribbean and decided to remain for a few months to recover his health while his servant, Hill, and most of their collections completed the journey to Britain as planned. A few months turned into nearly four years in the West Indies. The collection that Smeathman sent on ahead of him was large, delicate, and in varying states of decay. It initially sat untended in Drury's basement, as his subscribers expected Smeathman's return any day. The basement's humidity damaged many of the delicate specimens and the collection's immense size eventually forced Drury to rent a room to store it, explaining that once the items were unpacked from their boxes "My House could not possibly contain one half." Without Smeathman present to oversee unpacking, Drury had to beg naturalists among his acquaintance to help with the immense task of sorting, evaluating, and valuing the collections.[67] As they unpacked, Drury and the naturalists who assisted him discovered that "however hard & compact" termite nests were when they had first been collected, "they are now so fragile & tender that every one of them are broke to pieces ... & they all mould ... with amazing rapidity." Further, most of Smeathman's shells and birds were ruined from the way in which they had been preserved. Smeathman's plant specimens and beloved insects fared much better than his birds, shells, and termite nests.[68]

The collections, experiments, and observations Smeathman undertook while in Sierra Leone provided the basis for the natural historical work he presented and published when he did eventually return to Britain. These included a study of marine animalcules undertaken on the *Fly* during the outward voyage to Africa, a detailed weather journal, an essay about African spiders, and a description of the harmattan winds that he presented to the Society for the Promotion of Natural History in the 1780s. Smeathman's termite essay was completed in London based on notes, observations, drawings, and specimens collected in Sierra Leone.[69] Enough of Smeathman's specimens survived their journey to Britain to aid his efforts to establish himself as an authority on Sierra Leone's natural world. The surviving collection included pieces of termite nests, royal chambers, a mature queen preserved in spirits, and termites of each caste and species. Smeathman's surviving collection also allowed him to reinvent himself as a natural historical lecturer. During the early 1780s, the naturalist offered lectures in London and Paris that were, by all accounts, engaging and entertaining descriptions of termites' behavior, life cycles, and habits, and of his own adventures investigating the "many other curious and almost incredible circumstances of these extraordinary insects." Smeathman's surviving termite specimens played a starring role in his lectures as props passed around for audience members to examine. The popularity of Smeathman's lectures reflected the broader eighteenth-century fashion for scientific and medical lectures as a form of enlightened entertainment.[70]

In his harmattan essay, Smeathman observed that "notwithstanding our long acquaintance with this part of the globe, we are still ignorant of its metereology." The same could be said for much of the region's natural history. As the naturalist continued, "a residence of four years on the Coast, enabled me to make some observations, which from their novelty and accuracy, may possibly be useful."[71] Smeathman's four-year stay on the Upper Guinea coast enabled him to engage in the sorts of patient, long-term research projects that were common among his counterparts in Europe. His essay on termites, for example, was the work of years rather than weeks. Unlike slave ship surgeons and captains who also collected specimens in West Africa, Smeathman had the time and inclination necessary to engage in extensive natural historical studies. And unlike most of the individuals who gathered natural historical specimens along the routes of the slave trade, Smeathman was a naturalist in his own right.

FROM FLYCATCHER TO SLAVE TRADER

Smeathman's pursuit of natural history in Sierra Leone was fundamentally entwined with the local slave trade. The naturalist's close association with slaving captains and traders reshaped his attitudes toward slavery and enslaved Africans during the four years he spent along the Upper Guinea coast. Smeathman began his expedition skeptical of the slave trade and sympathetic toward the captive Africans he first encountered on a slave ship. Over the course of his residency in West Africa, he came to absorb many of the attitudes of the slaving captains and traders with whom he socialized.

Smeathman's first visit to a slave ship occurred before he even set foot on African soil. The vessel on which Smeathman traveled from England to Sierra Leone, the *Fly*, anchored off the Îles de Los on its way to the Banana Islands. The *Fly* encountered four other oceangoing vessels at the Îles de Los as well as a handful of sloops and schooners engaged in more local trade. Smeathman joined a gathering of European slaving captains and traders on board the largest of the transatlantic vessels, the *Africa*. The *Africa* was owned by the Liverpool slave merchant Miles Barber and was nearly ready to sail for Grenada with a full cargo of enslaved Africans. Smeathman reported that there were 400 captive Africans on board the vessel and a further 80 onshore.[72] The naturalist recalled that even before boarding the slave ship, his "ears were struck at some distance, with a confused noise of human voices & the clanking of chains which nearer & on board affects a sensible being with inexpressible horror." Smeathman's horror carries throughout his description of the *Africa*. Although Smeathman suggested that many of the children and women on board "seemed rather chearful than otherwise," he emphasized that this was not universally the case. Smeathman highlighted, in particular, the sorrow of two young mothers with "infants yet at the breasts." The two women displayed "a lively thoug[h] a silent grief; such as I had never seen more strongly marked in the human face." With babes still in arms, Smeathman believed the women had more reason to be grieved than their compatriots, and were fully sensible of it. He movingly described their "inexpressible anguish" as they watched their young children play on the ship's deck. "I was absorbed in a thousand melancholy reflections & bore a very small part in the conversation," the naturalist concluded.[73]

In his first encounter with a slave ship, Smeathman was clearly moved by what he saw, heard, and experienced. His entire description is one of someone thrust into a foreign and shocking world. In this first encounter with captive Africans, the naturalist portrayed them as fully human but subjected to inhuman conditions. He represents Africans as worthy of empathy and the transatlantic slave ship as a place of horror. By so doing, Smeathman implicitly disputed assertions made by defenders of the trade that captive Africans were better off enslaved, were treated well, and were insensible to the emotional and physical trauma of what they experienced.

Although Smeathman frequently traveled, slept, and dined on board transatlantic slaving vessels in the four years that followed this first visit to a slave ship, he rarely mentioned the slave trade or the slave ships he frequently visited. His account of the *Africa* was one of only two descriptions of a slave ship he wrote during the four years he lived in West Africa. Smeathman's second account, which was composed in 1773 on board another slaving vessel called the *Africa*, focused on the sickness and death that characterized the slave ship.[74] The sick and dying on board the *Africa* in 1773 included Smeathman's botanical assistant, Berlin.

Smeathman's second description of a slave ship emphasized the constancy of labor alongside sickness, suffering, and death. He described the fruitless efforts of the "raw & ignorant" slave ship surgeons and the unremitting labor of the often ill crew, noting that "if a man is attacked with a fever, they give him a vomit & set him to work again." He recorded the groans of sick captives and that "two or three slaves [were] thrown over board every day dying of fever, flux, measles, worms all together." Alongside the sounds of the sick and the rattling of chains, Smeathman described the constant noises of crew members, each busily employed in their respective labors.[75] The naturalist depicted the slave ship as humming with the work of the crew, "here they are hoisting casks, boxes & bales: the tackles creaking & the carpenter & coopers hammering & opening. There an armourer rasping & filing & cleaning arms." Smeathman's second account retained the sympathy for enslaved Africans that characterized the first, but their suffering became merely the backdrop to the orderly work of the crew. Missing is the naturalist's palpable horror at the slave trade; in its place is a fascination with industry, work, and labor.[76]

After two years on the Upper Guinea coast spent largely in the company of slave traders, the naturalist's attitude toward the slave trade and slavery markedly changed. In the same year that he composed the second description of a slave ship, Smeathman began to argue in defense of slavery. In May of 1773 he declared to the ardent abolitionist Fothergill that he believed "it will be a meritorious act to send as many black gentry to our Plantations as I can." He explained to Drury, "I have considered the matter & my scruples in regard to the slave trade are vanished, I hope and believe on good grounds." Although he did not explain the grounds upon which his principles had changed, he acknowledged that Fothergill, "master of every argument against slavery," would likely think him "rash in having as it were challenged him to the field."[77]

The naturalist went even further when writing to a more sympathetic audience. Smeathman announced to Banks, who was tepid at best toward abolitionism, that he was thinking of "turning dealer in souls as well as a merchant of butterflies & neettles." Smeathman continued, joking that the newly arrived Berlin "may turn some of his studies under Dr. Linneus to an advantage in examining whether some specimens of the *Primates* here, will be likely to meet with an agreable reception from the collectors in our Colonies, for which, I will make him the usual allowance, five per cent."[78] Smeathman's comments to Banks illustrate that he had adopted the dehumanizing attitudes toward Africans that were common among the slave traders with whom he socialized. His pun turned on the parallel he drew between collecting enslaved Africans and collecting specimens by referring to captive Africans as "specimens of the *Primates*." It pointed to the commodification of both natural historical specimens and enslaved Africans. In particular, it highlighted the callous attitude toward enslaved Africans that Smeathman had adopted by 1773. Whereas two years earlier he had depicted captive Africans sympathetically, he now quipped that they were another species of primate on whom a 5 percent profit might be made.

CONCLUSION

Smeathman's pun tellingly pointed to the deep connections between natural history, slaving, and commerce in the British Atlantic World. The naturalist arrived in Sierra Leone in 1771 as a traveling naturalist supported by a group of British patrons. He discovered that the patronage underwriting his expedition did not make his collecting efforts independent of the local and transatlantic slave trade. Smeathman came to understand that the pursuit of natural history along the Upper Guinea coast in the late eighteenth century fundamentally relied upon the commerce in human beings. The naturalist employed the routes, social infrastructure, and personnel of the slave trade to gather, transport, and study West African flora and fauna. He began his journey attempting to distance himself from the slave traders and captains he met in Sierra Leone. In time Smeathman came to not only rely on their assistance but also follow in their footsteps when he became an agent for a Liverpool-based slaving merchant.

The pro-slavery sympathies that Smeathman developed during his years in Sierra Leone proved short-lived. In 1775 Smeathman crossed the Atlantic

Ocean on board a slaver carrying more than 300 captive Africans. The journey inspired him to compose one final description of a slave ship. This account painted a very different picture than those that preceded it. It shared with his previous description an interest in the orderly labor of the slave ship, signaled through the essay's title, "Oeconomy of a Slave Ship." Yet, as literary scholar Deirdre Coleman observed, the naturalist now employed such language ironically. Coleman concluded that Smeathman's firsthand observations of the middle passage had taught him that "clearly there is no benign natural order underlying slavery."[79] After experiencing the middle passage firsthand and spending four years observing plantation slavery in the Caribbean, Smeathman returned to Britain seemingly committed to the abolitionist cause. Even the termite lectures he gave in London in the early 1780s concluded with an impassioned speech against the slave trade, demonstrating once again the entangled histories of science and slavery in the Atlantic World.[80]

Smeathman returned to London intent on leading another expedition to Sierra Leone. Some of the schemes he contemplated led contemporaries and modern scholars to question the depth of his commitment to abolitionism. Smeathman's plans included one for an agricultural colony that would purchase enslaved Africans who would then be freed immediately but required to repay the cost of their purchase through their labor. In the words of one contemporary, "he would buy although he would not sell" enslaved Africans.[81] The naturalist came close to realizing his dream of returning to Africa during the last year of his life. That year Smeathman published a colonization proposal, *Plan of a Settlement to be made near Sierra Leone* (1786), and secured the Committee for Relief of the Black Poor's support for it. The committee in turn convinced the British Treasury to back the colonization of Sierra Leone, largely along the lines outlined by Smeathman, and to place the naturalist in charge. Smeathman, however, fell ill and died in London on July 1, 1786, before the expedition set sail.[82]

The colonization of Sierra Leone in many ways presaged the next chapter in the history of British colonial science. Abolitionists supported the effort to establish a British colony in Sierra Leone because they believed it would help to demonstrate that ending the transatlantic slave trade would not necessarily damage the British imperial economy. Instead, alternative ways could be found to exploit Africa to the benefit of British commerce. Abolitionists' arguments eventually won the day. The British slave trade legally ended in 1807, but the entwined histories of British science and colonialism continued.

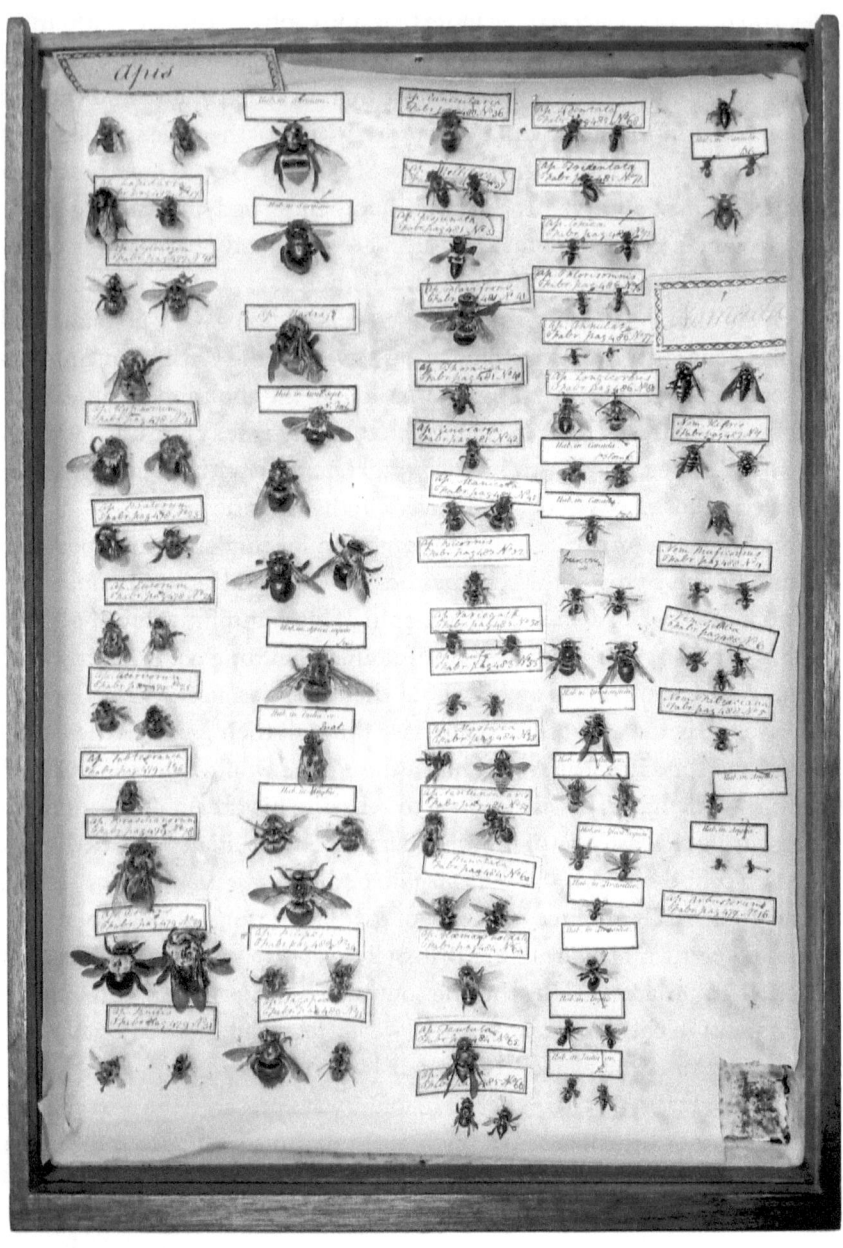

FIGURE 7.1. William Hunter's bee drawer. Hunter's original drawer displaying fifty-nine bee specimens, including four acquired by Henry Smeathman in Sierra Leone. GLAHM 150474–150581, Drawer HU/C/6, © The Hunterian, University of Glasgow, 2021.

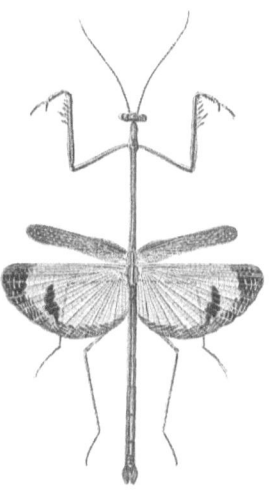

Epilogue

Among the many treasures of the Hunterian Museum at the University of Glasgow is William Hunter's original bee drawer (fig. 7.1). The glass-topped drawer from Hunter's cabinet is a stunning example of a late eighteenth-century natural historical display. The fifty-nine species of bees are arranged in four orderly columns, divided first by genera (*Apis* and *Nomada*) and then by species. Most species are represented by a pair, and all of the specimens are neatly held in place by brass pins. Eighteenth-century labels indicate specimens' scientific names or place of collection. The entire effect is one of orderly display.

Hunter's bee drawer seems at first glance to be a world away from the transatlantic slave trade. Yet four of the specimens in the drawer were collected by Henry Smeathman in Sierra Leone. They were likely transported on transatlantic slave ships and gathered by means of the social, political, and commercial infrastructure of the slave trade along the Upper Guinea coast. The four Sierra Leonean bees are part of a larger group of at least seventy-five entomological specimens gathered by Smeathman that can now be found in Glasgow. The Hunterian Museum originated with the bequest of the physician and anatomist William Hunter, who left his extensive collections to the University of Glasgow.[1] Hunter was not one of the sponsors of Smeathman's expedition to Sierra Leone. Thus, he did not receive any specimens directly from the naturalist. Understanding how Hunter's collections came to include at least seventy-five insects from Smeathman's expedition helps to illuminate the full scope of the impact the British slave trade had—and continue to have—on natural history and the production of natural knowledge.

There are at least three different ways that a Smeathman specimen could have joined Hunter's museum. Some of the Sierra Leonean insects in Hunter's collection are known to have come from Joseph Banks's museum through the intervention of the Danish entomologist Johan Christian Fabricius. Although Fabricius lived on the continent, he spent large stretches of time in London pursuing his research in the collections and libraries of British naturalists. The entomologist redistributed duplicate specimens among the collections where he worked, including those belonging to Hunter's and Smeathman's patrons. As a result, modern biologists and curators have located specimens (including syntypes) long after they were thought to have been lost and among collections where they did not expect to find them. Alternatively, Hunter may also have acquired Smeathman's specimens through the bequest or gift of one of Smeathman's patrons, such as Hunter's good friend John Fothergill.[2] Leaving one's museum to a more significant collector was one strategy for keeping a collection intact. Two of Smeathman's patrons—Drury and Margaret Cavendish Bentinck, Duchess of Portland—did not make similar arrangements. Consequently, their collections were sold at auction following their deaths. In both cases, thousands of specimens were divided into lots and sold over the course of multiple days to scores of collectors and specimen dealers. Shells, insects, and other specimens collected by Smeathman in Sierra Leone were redistributed to dozens of other collections through auctions.[3]

Similar stories of collections sold, bequeathed, gifted, traded, disbursed, and reorganized could be told for many of the natural objects gathered along the routes of the British slave trade. Early modern natural historical collections were not static. Individual specimens as well as entire museums changed hands and, in some cases, continued to do so decades later. Objects gathered by collecting slave traders therefore ended up in some unexpected places. Nearly 100 herbaria specimens prepared from plants gathered by the slave ship surgeon William Houstoun in New Spain and Jamaica are now owned by the state of California. They can be found today in San Francisco in the Sutro Library, part of the California State Library.[4] In this way institutions multiple degrees removed from the original collectors became the beneficiaries of the exploitations of the slave trade.

Slaving surgeons, slave ship mariners, and slave traders obtained and transported thousands of natural historical specimens by means of the British transatlantic slave trade over the course of the eighteenth century. The movement, disbursement, and destruction of specimens make it nearly impossible to fully quantify the scale of the collections gathered through the exploitations of the British slave trade. Yet the presence of Sierra Leonean bees in the Hunterian Museum and Houstoun herbaria specimens in the Sutro Library offer a hint of how wide and deep run the connections between natural history and the transatlantic slave trade.[5]

The abolition of the British slave trade in 1807 largely brought an end to British naturalists' reliance on slaving as a means of collecting. However, the slave trade was only one manifestation of the entwined histories of British science and colonialism. The entanglement of British science and colonialism intensified in the decades that followed abolition as the British Empire dramatically expanded over the course of the nineteenth century. Scientific and medical research enabled, and was enabled by, colonialism. British scientists relied on imperial officials, colonized peoples, sailors, soldiers, and others who traveled the empire's routes to gather the specimens, observations, and indigenous knowledges upon which natural inquiries were based. Scientific and medical breakthroughs both offered a justification for colonialism and solved some of the pragmatic, logistical, and epidemiological challenges of an empire upon which the sun never set. Furthermore, the British abolished only their slave trade in 1807; slavery itself continued in the British Empire for another three decades. Consequently, British naturalists in the early nineteenth century continued to rely directly and indirectly on slavery, to exploit Black bodies, and to attempt to marshal scientific arguments in defense of the slave economy.

Hunter's bees and Houstoun's pressed plants demonstrate that the history of natural history in the early modern Atlantic World cannot be told without reference to the history of the slave trade, slavery, and colonialism. The socioeconomic system that resulted in the brutal enslavement of millions of Africans also enabled the collection of thousands of plants, animals, shells, fossils, and other *naturalia*. These specimens circulated through the networks of the British slave trade and joined British herbaria, museums, and gardens. Their study resulted in new natural knowledge, including knowledge of how to further exploit Africa through the identification of new natural commodities and drugs. The specimens gathered by slave ship surgeons, slave traders, and others engaged in the slave trade gave British naturalists access to specimens that otherwise would have been difficult or impossible for them to acquire. Moreover, the subsequent redistribution, illustration, and publication of specimens gathered by slave traders amplified their impact within natural history. These publications, along with the specimens upon which they were based, should be accounted among the profits extracted by means of the slave trade.

Thinking broadly about the nature of slavery's profits can help to reframe modern conversations about the structural legacies of slavery. Public interest in understanding how today's racial inequities are rooted in the histories of slavery and colonialism has perhaps never been greater than in the wake of the Black Lives Matter movement. Growing public interest in understanding and confronting systemic inequality and racial injustice has reinvigorated efforts to address the legacies of slavery and colonialism. Among these efforts is the question of reparations. Although the idea of reparations has a long history—dating itself to the eighteenth century—it has been revisited in recent years with new urgency. Debates about reparations often hinge on the question of who benefited from slavery and the slave trade.[6] How might this question look if we expand our notion of slavery's profits? What if we measured slavery's profits in specimens collected and scientific papers published as well as in dollars and cents? If, instead of framing the question wholly in economic terms, we accounted the development of modern scientific knowledge and practice among slavery's profits, does this reshape how we formulate the answer?

Most natural historical objects gathered by means of the British slave trade do not survive to the present day. Some would have been lost or destroyed within days or months of their collection. Others would have fallen victim to the processes of natural decay or neglect over the intervening decades. Wars, natural disasters, disinterested heirs, and a host of other fates explain the destruction of still others. However, hundreds of specimens collected through the slave trade *do* survive in modern scientific institutions. This is particularly true of the more stable entomological and herbaria specimens. Plants, insects, shells, fossils, and other natural objects gathered along the routes of the slave trade are not merely of historical interest. Some continue to be used in modern research in biological systematics, ethnobotany, genetics, ecology, biodiversity, and historic climate change. Specimens gathered by slave traders include type specimens, the physical specimen that defines a species and against which other examples of the species must be compared. Historical collections provide a baseline against which scientists can assess the survival and distribution of species and knowledge of how to use them. Plants, insects, shells, fossils, and other *naturalia* collected through the routes of the slave trade thus continue to shape the production of natural knowledge.

The presence of specimens gathered by means of the slave trade raises questions for the institutions that own them and the scientists whose research requires their consultation. Questions of ethical use and display of collections are not new to museums. Institutions that hold ethnographic collections, especially human remains, have seen growing calls in recent decades for repatriation and, more broadly, calls to decolonize their practices and displays.[7] The origins of many modern museums lie in the colonial project, and colonialism frequently facilitated the acquisition of additional collections. The manner in which such objects traditionally have been displayed—for example, African art in ethnographic museums rather than in museums of art—perpetuated the idea that the cultures that produced the objects displayed exist only in the past rather than comprising a vibrant part of our present. Collections of skulls and skeletons, especially of Indigenous and African peoples, historically were used to develop and perpetuate pseudoscientific theories of racial difference and the creation of racial hierarchies.[8] In the United States, collections of human remains include individuals of African descent who had been enslaved. In response to pressure from activists, descendants, Indigenous communities, and governments, some nations and institutions have begun efforts to repatriate human remains for reburial and to repatriate art, objects, and other cultural resources.[9] Museums have also sought to address their historical connections to colonialism by reframing and replacing displays, revising the language employed internally and with visitors, and including communities whose cultural heritage they seek to interpret as partners in exhibit development. Conversations about decolonizing museums have rarely concerned floral, faunal, and mineral collections largely because the connections linking natural historical specimens to the histories of colonialism and slavery have been not well or widely understood.

How should plants, insects, shells, fossils, and other natural historical specimens collected by means of the slave trade be curated, displayed, and utilized? Natural history museums and other scientific institutions are beginning to engage in projects to better understand and address the historical links between their botanical, zoological, and mineral collections and slavery, the slave trade, and colonialism. Some have responded to calls to "decolonize science" with new interpretive programs, new research into the colonial histories of their collections, and new efforts to provide greater access to specimens obtained by means of colonialism, especially for scientists from the regions where the specimens were originally collected. Calls to decolonize science are best understood not as an attack on science but as an opportunity to make it more inclusive. By failing to acknowledge the ways the history of science is interwoven with the histories of slavery and colonialism, we risk contributing to public distrust of science, especially among the descendants of those exploited. Revealing—in all their difficult complexities—the means by which scientific knowledge was produced enriches the public's understanding of and trust in science today.[10]

By their very design, eighteenth-century natural historical displays abstracted specimens from their original contexts. Hunter's bee drawer, for example, placed specimens from Brazil, Canada, England, and Sierra Leone side by side to facilitate the comparison of species from across the globe. These displays made invisible the cultural, environmental, social, and historical contexts from which and by which specimens were collected. Such a display suited the needs of Hunter and his fellow eighteenth-century naturalists. The needs of our own historical moment are quite different. Finding ways to acknowledge how the slave trade formed one of the foundations of modern science would be a good place to start.

NOTES

ABBREVIATIONS

BL	British Library, London
Drury Catalogue	"A catalogue to exotic insects in the collection of Dru Drury," 1784, Dru Drury (1725–1803) Collection, Oxford University Museum of Natural History, England
Drury Letter Book	Letter-book of Dru Drury 1761–83, SB f D.6, Natural History Museum, London
HL	Huntington Library, San Marino, California
LLS	Library, Linnean Society of London, London
NHM	Natural History Museum, London
RS	Royal Society Library and Archives, London
TNA	National Archives of the United Kingdom, Richmond
Toller, "History of a Voyage"	William Toller, "The History of a Voyage to the River of Plate & Buenos Ayres From England," (1715), Mss 3039, Biblioteca Nacional de España, Madrid, Spain
UGSC	University of Glasgow Special Collections, University of Glasgow, Scotland
UUL	[Contemporary mss copy], MS D.26, Uppsala University Library, Uppsala University, Sweden

INTRODUCTION

1. John Burnet to unknown addressee, May 14, 1716, Sloane 4065, folio 248r, BL; "List of plants received from John Burnett," Sloane 4072, folio 295r, BL; and James Petiver, "The Following Curiosities were Presented me by my Hearty Friend Mr. John Burnet Surgeon to our English Factory at Porto Bello," Sloane 3331, folio 661r, BL.

2. *Slave Voyages: Trans-Atlantic* (voyage ID 76318); and South Sea Company, "MINUTES of the Court of Directors of the Governor and Company of Merchants of Great Britain Trading to the South Seas and other Parts of America and For Encouraging the Fishery," Add. MS 25496, vol. 3, 59, 143, 189, 202, 222, BL.

3. Hans Sloane, *Insects*, vol. 2, 180, Entomology Library, NHM.

4. "List of plants received from John Burnett," Sloane 4072, folio 295r, BL.

5. John Burnet to unknown addressee, May 14, 1716, Sloane 4065, folio 248r, BL; and "List of plants received from John Burnett," Sloane 4072, folio 295r, BL.

6. James Douglas, "The Description and Natural History of the Animal called Armadillo or the hog in armour from South America or the little American hog in Armour, by J. D.," MS Hunter D516, UGSC; James Petiver to James Douglas, March 30, 1716, MS Hunter D513, UGSC; and Murphy, "A Slaving Surgeon's Collection."

7. "List of plants received from John Burnett," Sloane 4072, folio 295r, BL.

8. "List of plants received from John Burnett," Sloane 4072, folio 295r, BL; and James Petiver to William Toller, November 19, 1716, Sloane MS 3340, folios 275v–276r, BL.

9. Richardson, "The British Empire and the Atlantic Slave Trade," 440.

10. Eric Herschthal's fascinating recent book demonstrates how during the late eighteenth and early nineteenth centuries antislavery activists, including scientists, in Britain and the United States employed scientific arguments in support of the abolitionist cause. Herschthal, *Science of Abolition*.

11. Terrall, *Catching Nature in the Act*, 1–7, 13–43.

12. Cook, "The Cutting Edge of a Revolution?," 58 (quotation), 54.

13. Delbourgo, *Collecting the World*.

14. Lucas and Lucas, "Natural History 'Collectors.'"

15. Cook, *Matters of Exchange*; and Margócsy, *Commercial Visions*. See also Dorner, *Merchants of Medicines*; Pratt, *Imperial Eyes*; Schiebinger and Swan, *Colonial Botany*; and Stewart, "Global Pillage."

16. As detailed below, there is a dynamic and growing body of scholarship focused on the broader question of the relationship between slavery and natural inquiry in the early modern Atlantic World. Much less work has been done on the narrower question of how the infrastructure of the transatlantic slave trade shaped scientific and medical inquiries. Manuel Barcia's recent book is a notable exception to this trend. However, Barcia's thematic and temporal focus are distinct from mine. Barcia focuses on British medicine and medical knowledge in the nineteenth century after the British outlawed the transatlantic slave trade, whereas *Captivity's Collections* focuses on British natural history during the eighteenth century before the rise of the abolition movement. Barcia, *Yellow Demon of Fever*. Other examples of work examining the relationship between the infrastructure of the slave trade and natural inquiry include Dew, "Scientific Travel in the Atlantic World"; Douglas, "Making of Scientific Knowledge"; and Schaffer, "Golden Means."

17. Golinski, *Making Natural Knowledge*, 2 (quotation), 2–3.

18. Herschthal, *Science of Abolition*, 247 (quotation), see esp. 2–16, 246–52.

19. See, for example, Baptist, *Half Has Never Been Told*; Baucom, *Specters of the Atlantic*; Beckert, *Empire of Cotton*; Beckert and Rockman, *Slavery's Capitalism*; Johnson, *River of Dark Dreams*; Pettigrew, *Freedom's Debt*; Schermerhorn, *Business of Slavery*; and Wilder, *Ebony and Ivy*.

20. The Legacies of British Slave-ownership project substantiated many of Eric Williams's pioneering claims about the impact of wealth derived from slavery on British culture and society. Williams, *Capitalism and Slavery*.

21. Hall et al., *Legacies of British Slave-ownership*; and Draper, *Price of Emancipation*.

22. Huxtable et al., "Interim Report." British art museums like the National Gallery have undertaken similar investigations into the connections between their collections and British slavery. See "National Gallery and Legacies of British Slave-ownership Research Project," November 8, 2021, www.nationalgallery.org.uk/research/research-partnerships/legacies-of-british-slave-ownership-research-project. The Natural History Museum in London was an early leader in drawing attention to slavery's connections with the pursuit of natural history. The museum put on a series of public events and a temporary exhibit in its Library's Rare Books Room as part of the national commemoration in 2007 of the bicentenary of the abolition of the British slave trade. For more information, see Tracy-Ann Smith et al., "Slavery and the Natural World," Natural History Museum, London (n.d.), www.nhm.ac.uk/discover/slavery-and-the-natural-world.html.

23. Williams, *Capitalism and Slavery*. Other aspects of Williams's argument not directly relevant here, including the contention that plantation slavery in the British Caribbean was in decline by the late eighteenth century, have also been the subject of extensive debate. Recent scholarship on slavery and capitalism has deepened our understanding of the ways in which slavery was foundational to the development of capitalism. This recent scholarship, much of which focuses on the United States, builds on many of the observations by Williams as well as by W. E. B. Du Bois, C. L. R. James, and Cedric Robinson. See, for example, Baptist, *Half Has Never Been Told*; Beckert, *Empire of Cotton*; Beckert and Rockman, *Slavery's Capitalism*; and Schermerhorn, *Business of Slavery*.

24. McClellan, *Colonialism and Science*; and Delbourgo, *Collecting the World*.

25. Iannini, *Fatal Revolutions*; and Kriz, "Curiosities, Commodities, and Transplanted Bodies."

26. Barcia, *Yellow Demon of Fever*; Breen, *Age of Intoxication*; Cagle, *Assembling the Tropics*; Chakrabarti, *Materials and Medicine*; Curran, *Anatomy of Blackness*; Dorner, *Merchants of Medicines*; Fett, *Working Cures*; Hogarth, *Medicalizing Blackness*; Seth, *Difference and Disease*; Walker, "The Medicines Trade"; and Weaver, *Medical Revolutionaries*.

27. Delbourgo, *Collecting the World*; Gómez, *Experiential Caribbean*; Lambert, *Mastering the Niger*; Murphy, "Translating the Vernacular"; Ogborn, "Talking Plants"; Parrish, *American Curiosity*; Schiebinger, *Plants and Empire*; and Schiebinger, *Secret Cures of Slaves*.

28. The diffusionist model is most associated with George Basalla. See Basalla, "The Spread of Western Science." Recent work in science and empire is extensive. See, for a start, Bleichmar, *Visible Empire*; Cagle, *Assembling the Tropics*; Chakrabarti, *Materials and Medicine*; Dorner, *Merchants of Medicines*; Gómez, *Experiential Caribbean*; Parrish, *American Curiosity*; Raj, *Relocating Modern Science*; Schiebinger, *Plants and Empire*; Schiebinger, *Secret Cures of Slaves*; and Seth, *Difference and Disease*.

29. Raj, *Relocating Modern Science*, 20.

30. As described in chapter 3, Burnet became a spy for the Spanish Crown and provided evidence against the British South Sea Company. Spanish sources establishing Burnet's credentials and credibility fill in some important gaps in his personal narrative, including providing the name of the slave ship on which he first sailed to Spanish territories, the *Wiltshire*. Brown, "The South Sea Company," 670.

31. Petiver, *Gazophylacii Naturae & Artis, Decas Prima*, Tab. VI, folio 9r–v.

32. Savitt, "The Use of Blacks"; Fabian, *Skull Collectors*; and Kenny, "Development of Medical Museums."

33. Although I expected to find other examples in the course of my research, the *Wiltshire* collection is the only one for which I have found clear evidence of the inclusion of human

remains. The limited evidence of human specimens among the collections of slave traders likely reflects the fragmentary nature of the sources as well as my focus on natural history rather than the history of medicine.

34. Chakrabarti, *Materials and Medicine*, 11.

35. The phrase "slavery at sea" comes from Sowande' Mustakeem's 2016 book, which she describes as part of the small but growing middle passage studies scholarship. Mustakeem, *Slavery at Sea*, 4. See also Christopher, *Slave Ship Sailors*; Rediker, *Slave Ship*; and Smallwood, *Saltwater Slavery*.

CHAPTER ONE

1. *Slave Voyages: Trans-Atlantic* (voyage ID 15235); Christopher, *Slave Ship Sailors*, 33; Rediker, *Slave Ship*, 4, 59, 212, 237; and Sheridan, "Guinea Surgeons," 615–16.

2. James Petiver to Robert Barcklay, ca. 1710, Sloane MS 3337, folios 109v–110r, BL.

3. Technically, the slave trade was English, rather than British, before the 1707 Acts of Union. Throughout this book "English" is used when referring exclusively to the period before 1707, and "British" is used when referencing events that occurred after 1707, including those that began before 1707.

4. The natural historical specimens collected in Africa by means of the British slave trade in the early eighteenth century were overwhelmingly from West Africa, although a handful came from West Central Africa. Petiver, for example, acquired only five specimens from West Central Africa, all of which he described as Angolan.

5. James Petiver to James Fraser, January 27, 1707, Sloane 3335, folio 48v, BL.

6. Shapin, *Social History of Truth*, 355–408; Murphy, "Translating the Vernacular"; and Parrish, *American Curiosity*, 259–306.

7. James Petiver to John Smyth, February 21, 1693, Sloane 3332, folio 33r, BL; James Petiver to Dr. [John] Smyth, n.d. [1694?], Sloane 3332, folios 166v–167v, BL; Petiver, "Catalogue of some *Guinea-Plants*," 677–86; and Sloane Herbarium, H.S. 191, NHM. Petiver's "Catalogue of some *Guinea-Plants*" listed forty specimens. The corresponding herbaria specimens suggest the collection originally included forty-six specimens, forty-four of which survive today in the Sloane Herbarium. Dandy, *Sloane Herbarium*, 53; and Soelberg et al., "Historical Versus Contemporary Medicinal Plant Uses," 110, 114.

8. Davies, *Royal African Company*, 1–44; Hair and Law, "The English in Western Africa"; and Pettigrew, *Freedom's Debt*, 22–24.

9. Hair and Law, "English in Western Africa," 257–62; Daaku, *Trade and Politics*, esp. 48–72; Davies, *Royal African Company*; and Newman, *A New World of Labor*, 110.

10. Shumway, *The Fante*, 48 (quotation), 22–56; and Schaffer, "Golden Means."

11. The region between the Senegal and Sierra Leone Rivers was initially called Guinea, but by the second half of the seventeenth century, Guinea was used in England to refer to the entirety of West Africa.

12. October 31, 1694, Journal Books of Scientific Meetings, Collections from the Royal Society, 1660–1800, vol. 8, 259. See also March 20, 1695, Journal Books of Scientific Meetings, Collections from the Royal Society, 1660–1800, vol. 8, 286; Grew, *Musaeum Regalis Societatis*, 31; Porter, "The Crispe Family and the African Trade," 74–75; Classified Papers, vol. 19, "Inquiries and Answers, December 24, 1662–May 25, 1692," The Early Letters and Classified Papers, 1660–1740, Collections from the Royal Society, no. 74; Govier, "The Royal Society, Slavery and the Island of Jamaica," 214; Hall, *Slavery and African Ethnicities*, 80–81; and

J. Hillyer to Dr. Bathurst, January 3, 1687 and April 25, 1688, The Early Letters and Classified Papers, 1660–1740, Collections from the Royal Society, no. 76.

13. Petiver, "Account of a Book," 399. See also Jarvis, "'Most Common Grass,'" 316.

14. Stearns, "James Petiver"; Murphy, "Collecting Slave Traders"; and Swann, *Curiosities and Texts*, 90–96.

15. Royal African Company, October 4, 1692, "Letters to Guinea/Letters Sent," T70/50, folio 139r, TNA; August 27, 1703, "Extracts of Letters received by the Royal African Company of England so far as relate to the Committee of Correspondence, No. 4, From February the 13th 1702 To May the 9th 1704," folio 47r, T70/13, TNA; Daaku, *Trade and Politics*, 98–99; and Law, *English in West Africa*, 30n132.

16. Law, *English in West Africa*, 29–30, 49, 122–24, 413–16, 565–67.

17. Daaku, *Trade and Politics*, 98–99; and Bosman, *A New and Accurate DESCRIPTION*, 51 (quotation), 51–52.

18. James Petiver to Edward Bartar, February 27, 1693, Sloane 3332, folio 33v, BL. Leonard Plukenet was the royal professor of botany and gardener to Queen Mary. Samuel Doody was an apothecary and keeper of the Apothecary's Garden in Chelsea. Both were likely members of the Temple Coffee House botany club. Petiver may have been referencing this group when he wrote that they drank to Bartar's memory each week. Riley, "The Club at the Temple Coffee House."

19. J[ohn] Smyth to Edward Bartar, October 10, 1696, Sloane MS 3332, folio 250r, BL.

20. James Petiver to Edward Bartar, October 15, 1694, Sloane 3332, folio 164v, BL.

21. James Petiver to Edward Bartar, April 24, [1694], Sloane MS 3332, folios 55v–56v, BL.

22. James Petiver to Edward Bartar, October 28, 1694, Sloane 3332, folio 84r–v, BL. Petiver similarly recommended that his correspondents in the Caribbean employ enslaved collectors. See Murphy, "Petiver's 'Kind Friends,'" 14.

23. James Petiver to Edward Bartar, October 4, 1697, Sloane MS 3333, folio 69v, BL.

24. Delbourgo, *Collecting the World*, 94–100; Gómez, *Experiential Caribbean*; Parrish, *American Curiosity*, 259–306; Safier, *Measuring the New World*, 62–66; and Schiebinger, *Plants and Empire*, 73–104.

25. Petiver, *Musei Petiveriani, Centuria Prima*, 7, 9; and Petiver, *Musei Petiveriani, Centuria Secunda & Tertia*, 22.

26. J[ohn] Smyth to Edward Bartar, October 10, 1696, Sloane MS 3332, folio 250r, BL; James Petiver, "A Description of divers Animals, Shells, Insects, Plants, &c lately Observed on the Coasts of Guinea with their Figures Communication by J. P. Apoth. F. R. S.," Sloane 1968, folio 166r, BL; and James Petiver to Edward Bartar, October 28, 1694, Sloane 3332, folio 84r, BL.

27. Petiver, *Musei Petiveriani, Centuria Quarta & Quinta*, 43; Petiver, *Gazophylacii Naturae & Artis, Decas Septima and Octava*, tab. 69; and Petiver, "A Description of divers Animals Shells, Insects, Plants, &c lately Observed on the Coasts of Guinea with their Figures Communication by J. P. Apoth. F. R. S.," Sloane 1968, folio 166r, BL.

28. Sloane Herbarium, HS 155, folio 161r–v, NHM. The label is not in Petiver's distinctive handwriting. It is also not an obvious match for other known contributors to the Sloane Herbarium. This, combined with Petiver's habit of keeping a collector's original label with the specimen, raises the strong possibility that the label was written by Bartar. My many thanks to Dr. Mark Carine of the Natural History Museum of London for his assistance with and expertise about Bartar's specimens in the Sloane Herbarium. For Petiver's tendency to keep the original labels, see Jarvis, "'The Most Common Grass,'" 312.

29. James Petiver to James Fraser, January 27, 1707, Sloane 3335, folio 48v, BL; and James Fraser to James Petiver, March 17, 1707, Sloane 3321, folio 215r–v, BL. For Fraser's earlier voyage, see *Slave Voyages: Trans-Atlantic* (voyage ID 21162); James Fraser to James Petiver, November 14, 1703, Sloane 3321, folio 130r–v, BL. For slave ship surgeons' chests, see Taylor, "Sea-Surgeons," 101; and Sheridan, "Guinea Surgeons," esp. 611–12. For the idea that Petiver would remember his correspondents "to posterity" through printed acknowledgment, see, for example, James Petiver to Edward Bartar, February 27, 1693, Sloane 3332, folio 33v, BL.

30. James Fraser to James Petiver, March 17, 1707, Sloane 3321, folio 215r–v, BL.

31. James Petiver to James Fraser, January 27, 1707, Sloane 3335, folio 48v, BL. For the texts referenced in this letter, see Petiver, *Gazophylacii Naturae & Artis, Deca Prima*; and Petiver, "Catalogue of some *Guinea-Plants*," 677–86.

32. Petiver, *Gazophylacii Naturae & Artis, Deca Prima*, 5, 9–11, 14–15.

33. William Pettigrew identifies three of the four owners of the *Mayflower* as independent traders. Pettigrew, "Directory of Independent Slave Traders," in *Freedom's Debt*, 229, 232–33. For the owners of the *Mayflower*, see *Slave Voyages: Trans-Atlantic* (voyage ID 21162).

34. Richardson, "The British Empire," 445.

35. Pettigrew, *Freedom's Debt*, 11.

36. Pettigrew, *Freedom's Debt*, 4–7, 11–16, 22–44, 155–59; Davies, *Royal African Company*, 44–46, 97–152; and Richardson, "The British Empire," 444–46.

37. Hans Sloane, Manuscript Catalogues of Sir Hans Sloane's Collections: Echini Marini (Zoology), n.p., MSS SLO, NHM; Hans Sloane, Manuscript Catalogues of Sir Hans Sloane's Collections: Fossils, vols 1–5, vol. 5, folio 367r–v (quotation), Paleontology Library, MSS SLO, NHM; Hans Sloane, Manuscript Catalogues of Sir Hans Sloane's Collections: Miscellanies Catalogue, items 1257–60, 1829, Centre of Anthropology, Department of Africa, Oceania, and the Americas, British Museum; Hans Sloane, Manuscript Catalogues of Sir Hans Sloane's Collections: Index to *Vegetable and Vegetable Substances: being the original register of the plant collections of Sir Hans Sloane excluding the Herbarium, arranged in the order of their acquisition*, vol. 3, folio 983r–v (quotation), NHM; and Hans Sloane, Manuscript Catalogues of Sir Hans Sloane's Collections: Insects, MSS SLO, vols. 1–2, NHM.

38. Hans Sloane, Manuscript Catalogues of Sir Hans Sloane's Collections: Miscellanies Catalogue, p. 181, item 1368, Centre of Anthropology, Department of Africa, Oceania, and the Americas, British Museum; Delbourgo, *Collecting the World*, 230; and MacGregor, *History of the World*, 560–65.

39. James Douglas, "A botanical dissection of the fruit of the coco tree," MS Hunter D418, UGSC; and James Douglas, "Fructus Coco," MS Hunter D419, UGSC.

40. Twenty of the thirty-seven mariners (68 percent) in Petiver's Atlantic network were engaged in the slave trade. Murphy, "Petiver's 'Kind Friends,'" 262, 267–73.

41. Petiver, *Gazophylacii Naturae & Artis, Decas Prima & Secunda*, 5, 9, 15, 19; Petiver, *Musei Petiveriani Centuria Nona & Decima*, 87, 89, 95; Petiver, "A Description of divers Animals," folio 166v, BL; and Sloane, *Insects*, vol. 2, 290–91, NHM. After Watts's death, Skeen added Watts's specimens to the collection he eventually presented to Petiver.

42. Petiver, *Gazophylacii Naturae & Artis, Decas Prima & Secunda*, 5, 11, 23 ("harmless and very beautiful"); Petiver, *Gazophylacii Naturae & Artis, Decas Tertia*, 44 ("the *Guinea Coast*"); Petiver, *Musei Petiveriani: Centuria Quarta & Quinta*, 46 ("very curious"), 49, 51, 52, 55; and Petiver, *Musei Petiveiani: Centuria Nona & Decima*, 88, 95 ("not yet"). See also Petiver, *Musei Petiveriani: Centuria Secunda & Tertia*, 24, 46.

43. James Petiver to Edward Bartar, October 4, 1697, Sloane MS 3333, folio 69v, BL; and James Petiver to Edward Bartar, November 18, 1698, Sloane MS 3333, folios 236v–237r, BL.

44. Davis, *Inhuman Bondage*, 88–91; Eltis, *Rise of African Slavery*, 2, 137–49, 164; and Thornton, *Africa and Africans*.

45. Parsons and Murphy, "Ecosystems under Sail"; and Rigby, "Politics and Pragmatics."

46. James Petiver to Edward Bartar, February 20, 1697, Sloane MS 3332, folio 251r–v, BL; and James Petiver to Edward Bartar, November 18, 1698, Sloane MS 3333, folios 236v–237r, BL. For the *Prince George*, see *Slave Voyages: Trans-Atlantic* (voyage ID 9722).

47. David Crawford to James Petiver, September 25, 1706, Sloane 3321, folio 234r–v, BL; James Petiver to David Crawford, February 3, 1707, Sloane 3335, folios 49r–v, BL; James Petiver to David Crawford, February 14, 1707, Sloane 3336, folio 18r–v, BL; James Petiver to Mr. Lancett, February 16, 1707, Sloane 3336, folio 19r–v, BL; James Petiver to David Crawford, December 2, 1711, Sloane 3337, folio 158r–v, BL; and David Crawford to James Petiver, May 17, 1712, Sloane 3321, folio 279r–v, BL.

48. James Petiver, "A Catalogue of such Persons to Whom I have given my first Century," Sloane 3332, folio 161v, BL; and James Petiver to James Fraser, January 27, 1707, Sloane 3335, folio 48v, BL.

49. James Fraser to James Petiver, March 17, 1706, Sloane MS 3321, folio 215r–v, BL; and James Fraser to James Petiver, November 14, 1703, Sloane MS 3321, folio 130r–v, BL. The *Slave Voyages: Trans-Atlantic* database shows that the *Mayflower* departed London on September 8, 1703. Fraser wrote Petiver from Plymouth two months later, noting that the *Mayflower* might not depart until Christmas. The ship's crew began purchasing captive Africans in Whydah in mid-March 1704. *Slave Voyages: Trans-Atlantic* (voyage ID 21162).

50. Most likely Petiver's initial negotiations with potential maritime collectors were conducted face-to-face. Most of the surviving initial letters between Petiver and his maritime collectors were framed as reminders of promises previously agreed upon.

51. For gifts of books, food, or drink, see James Petiver to David Patton, July 8, 1717, Sloane MS 3340, folios 330v–331r, BL; and James Petiver to Edward Bartar, February 20, 1697, Sloane MS 3332, folio 251r, BL. For gifts of medicines, see Petiver to Edward Bartar, October 15, 1694, Sloane MS 3332, folio 166r, BL; Petiver to Hannah Williams, November 17, 1706, Sloane MS 3335, folio 40v, BL; and Edmond Bohun to Petiver, April 20, 1700, Sloane MS 3321, folio 40r, BL. For Petiver's offers to pay per specimen, see Petiver to David Crawford, December 2, 1711, Sloane MS 3337, folio 158r, BL; and Petiver to Mr. Rickets, ca. 1708, Sloane MS 3336, folio 40v, BL. See also Stearns, "James Petiver," 264.

52. Sloane, *Vegetable and Vegetable Substances*, vol. 3, p. 915, item 7627, NHM. For the African souvenir market, see Webster, "Material Culture of Slave Shipping," esp. 113–15.

53. Aubry, *The Sea-Surgeon*, 118–20; Sheridan, "Guinea Surgeons," 615–16; and Rediker, *Slave Ship*, 4, 59, 212, 265.

54. Parsons and Murphy, "Ecosystems under Sail," 522–30.

55. James Petiver to George Jesson, July 20, 1716, Sloane MS 3340, folio 252r, BL. Alternatively, Petiver may have intended that Jesson employ free or enslaved Black sailors to gather specimens.

56. Murphy, "Collecting Slave Traders," 648n25.

57. Rediker, *Slave Ship*, 57, 59. See also Christopher, *Slave Ship Sailors*, 33; and Watson, "The Guinea Trade," 203–14, esp. 206. Although slave ships were the only British merchant ships in the Atlantic that typically employed surgeons in this period, surgeons were typically

included among the crews of the Royal Navy and of British merchant ships engaged in long-distance trade, such as to the East Indies.

58. Maydom, "James Petiver's Apothecary Practice."

59. This is not to suggest that slave ship surgeons were typically well educated, simply that they on average were better trained regarding natural inquiries than the rest of a slave ship's crew. Slave ship surgeons had a reputation for being poorly trained, which modern scholarship has largely substantiated. Stephen D. Behrendt's research on slave ship surgeons in the late eighteenth century indicates that royal naval impressment officers viewed slave ship surgeons as less qualified than their naval counterparts. This partially explains the lower rate at which they were impressed into the Royal Navy. When slave ship surgeons were impressed, they were generally assigned the lower status of acting assistant surgeon or surgeon's mate rather than the rank of ship surgeon. Behrendt, "The Captains," 120n61.

60. James Sutherland to James Petiver, March 25, 1700, Sloane MS 4063, folios 9r–10v, BL; and Sutherland to Petiver, June 24, 1700, Sloane MS 4063, folio 32r–v, BL.

61. For ship surgeons' education and training, see Taylor, "Sea-Surgeons," 101; Sheridan, "Guinea Surgeons," 611–12; Watson, "The Guinea Trade," 211–13; and Chamberlain, "Influence of the Slave Trade," 771. For physicians and natural history, see Cook, "Physicians and Natural History," 91–105.

62. Sheridan, "Guinea Surgeons," 616 (quotation), 615–16; Rediker, *Slave Ship*, 4, 59, 212, 265; and Aubry, *Sea-Surgeon*, 118–20.

63. Daston, "Type Specimens," 166.

64. James Petiver to Dr. [John] Smyth, n.d. [1694 or 1695], Sloane 3332, folios 166v–167v, BL. Petiver entrusted the letter to George Winchfield, surgeon on the Royal African Company's vessel, *Kendall*. See *Slave Voyages: Trans-Atlantic* (voyage ID 9725).

65. December 21, 1698, January 24, 1700, and November 13, 1700, Journal Books of Scientific Meetings, Collections from the Royal Society, 1660–1800, vol. 9, 118, 190, 229; and Petiver, "A Description of divers Animals Shells, Insects, Plants, &c lately Observed on the Coasts of Guinea with their Figures Communication by J. P. Apoth. F. R. S.," Sloane 1968, folio 166r, BL. For Forty, see *Slave Voyages: Trans-Atlantic* (voyage ID 20229).

66. For Lister, see Roos, *Web of Nature*, 309–10; and James Petiver to Edward Bartar, October 28, 1694, Sloane 3332, 85r, BL. For Plukenet, see Leonard Plukenet, *Almagestum Botanicum*, tab. 276; and Austin, "Sendera-clandi," 442. For Sherard, see Clokie, *Account of the Herbaria*, 76. For Dillenius, see Johann Jacob Dillenius, "Plantae Petiv. ex Artis Philosoph. quam nomina Speciminibus nondum addita," MSS Sherard 205, folios 44r–48v, Sherardian Library, University of Oxford.

67. Soelberg et al., "Historical Versus Contemporary Medicinal Plant Uses."

68. Some of Petiver's contemporaries complained about "the poor quality of the engravings and the brevity of the descriptive text accompanying them" in his texts. See Hunt, "Under Sloane's Shadow," 209–10.

69. Petiver, *Gazophylacii Naturae & Artis: Decas Tertia*, tab. 29, fig. 3 (quotation), 44; Petiver, "A Description of divers Animals, Shells, Insects, Plants, &c lately Observed on the Coasts of Guinea with their Figures Communication by J. P. Apoth. F. R. S," Sloane MS 1968, folio 166r–v, BL; and Linnaeus, *Systema Naturae*, 1:496. For the importance of images to the practice of early modern natural history, see Bleichmar, *Visible Empire*, 43–72.

70. Petiver, "Catalogue of some *Guinea-Plants*," 677–86. Petiver concluded the catalog by noting he planned to present additional African plant specimens at future Royal Society

meetings and by declaring it was his "great Ambition to approve my self." Petiver, "Catalogue of some *Guinea-Plants*," 686.

71. Petiver, "Catalogue of some *Guinea-Plants*," 677–86.

72. Petiver, "Catalogue of some *Guinea-Plants*," 677; and Soelberg et al., "Historical Versus Contemporary Medicinal Plant Uses," 110–14, 117. Petiver's emphasis on African names and knowledges was the subject of a scathing satirical account of the text in William King's *The Transactioneer* (1700). See Murphy, "Petiver's 'Kind Friends,'" 274.

73. William Ryan's recent analysis of Petiver's "Catalogue of some *Guinea-Plants*" makes a compelling argument that Petiver used analogies, Latinate classifications, and passive voice in attempting to familiarize African flora and exert control over the text. In neither goal was Petiver fully successful. Ryan, "Imperfect Knowledge," 134–36.

74. Petiver, "Catalogue of some *Guinea-Plants*," 678 (quotation), 677–86; Dandy, *Sloane Herbarium*, 53; Jarvis, "'Most Common Grass,'" 312–27; and Soelberg et al., "Historical Versus Contemporary Medicinal Plant Uses," 109–14. Petiver's herbaria specimens typically mirrored his published descriptions of the plant. See Jarvis, "'Most Common Grass,'" 318.

75. For example, see James Petiver to Edward Bartar, October 4, 1697, Sloane MS 3333, folio 69v, BL. A *materia medica* was a text that described substances, usually derived from plants, that were known for their healing properties.

76. Breen, *Age of Intoxication*; Cagle, *Assembling the Tropics*; Dorner, *Merchants and Medicines*; and Walker, "Medicines Trade."

77. Curtin, "'The White Man's Grave.'"

78. Mortality rates for Europeans arriving in West Africa in the later eighteenth century were between 30 and 70 percent for the first year. Thereafter, morality rates ranged from 8 to 12 percent. In comparison, mortality in the seventeenth-century Chesapeake—also considered to be a particularly unhealthy place for Europeans—ranged between 5 and 7 percent. Captives and captors in West Africa suffered from a range of diseases, including dysentery, sleeping sickness, yaws, yellow fever, and malaria. The collective toll of these diseases helps to explain why the life expectancy for an officer of the Royal African Company in the early eighteenth century was between four and five years, while noncommissioned soldiers and artisans fared even worse. Those Royal African Company employees who survived were often incapacitated by illness for long stretches of time. Curtin, "'The White Man's Grave'"; and Newman, *New World of Labor*, 109, 114–15, 125–29.

79. Cagle, *Assembling the Tropics*; Chakrabarti, *Materials and Medicine*; Crawford and Gabriel, "Introduction"; Gómez, *Experiential Caribbean*; and Schiebinger, *Secret Cures of Slaves*.

80. Petiver, "Catalogue of some *Guinea-Plants*," 678, 682.

81. Petiver, "Catalogue of some *Guinea-Plants*," 678, 680.

82. Petiver, "Catalogue of some *Guinea-Plants*," 677 (quotations); Cook, *Matters of Exchange*, 207–8; Murphy, "Translating the Vernacular"; Schiebinger, *Plants and Empire*, 73–104; and Soelberg et al., "Historical Versus Contemporary Medicinal Plant Uses." My thanks to Claire Gherini for sharing her expertise in the history of medicine with me on this and many other points.

83. In addition to "Catalogue of some *Guinea-Plants*," Petiver also described Cape Coast specimens obtained from Smyth in *Musei Petiveriani, Centuria Prima*, 6; and Petiver, "Account of a Book."

84. Breen, "Flip Side," 146.

CHAPTER TWO

1. Hans Sloane, Manuscript Catalogues of Sir Hans Sloane's Collections: Index to *Vegetable and Vegetable Substances*, vol. 3, 956–60, NHM; James Brydges, Duke of Chandos to Capt. General Tinker, December 9, 1723, Stowe 57, vol. 23, folio 135r–v, HL; and *Slave Voyages: Trans-Atlantic* (voyage ID 75286).

2. Sloane, Index to *Vegetable and Vegetable Substances*, vol. 3, 956–60, NHM.

3. Chandos to Capt. General Tinker, December 9, 1723, Stowe 57, vol. 23, folio 135r–v, HL; and Chandos to Capt. General Tinker, December 21 and 29, 1723, Stowe 57, vol. 23, folios 176r–178v, HL.

4. Richardson, "The British Empire," 445.

5. Pettigrew, *Freedom's Debt*, 4–13, 37–44.

6. Chandos to James Phipps, July 21, 1721, Correspondence of the Duke of Chandos, C113/279, folio 12r, TNA.

7. Chandos to James Phipps, January 6, 1721, Stowe 57, vol. 17, folio 318r–v, HL.

8. Chandos to Ambrose Baldwyn, July 7, 1722, Stowe 57, vol. 21, folio 13r–v, HL; and Duke of Chandos to Hans Sloane, December 4, 1721, Sloane 4046, folio 152r, BL.

9. Chandos to Ambrose Baldwyn, July 7, 1722, Stowe 57, vol. 21, folio 13r–v, HL.

10. Baker and Baker, *Life and Circumstances*; Jenkins, *Portrait of a Patron*; Pettigrew, *Freedom's Debt*, esp. 165–72; and Stewart, *Rise of Public Science*, 312–26.

11. Stewart, *Rise of Public Science*, 214 (quotation), 312–26; and Pettigrew, *Freedom's Debt*, esp. 165–72.

12. Mitchell, "'Legitimate commerce,'" 556 (quotation); and Pettigrew, *Freedom's Debt*, 162–65. Mitchell argues that Chandos and his allies kept their plans secret from the majority of the company's investors and officials, who continued to believe the company would focus on the transatlantic slave trade.

13. For a start, see Barrera-Osorio, *Experiencing Nature*; Bleichmar, *Visible Empire*; Cook, *Matters of Exchange*; Chakrabarti, *Materials and Medicine*; Delbourgo and Dew, *Science and Empire*; Drayton, *Nature's Government*; McClellan, *Colonialism and Science*; MacLeod, *Nature and Empire*; Parrish, *American Curiosity*; Raj, *Relocating Modern Science*; Schiebinger, *Plants and Empire*; Schiebinger and Swan, *Colonial Botany*; and Winterbottom, *Hybrid Knowledge*.

14. For example, see Chandos to Phipps, September 15, 1721, Correspondence of the Duke of Chandos, C113/279, folio 27r–v, TNA.

15. Chandos to Phipps, July 21, 1721, Correspondence of the Duke of Chandos, C113/279, folios 12v–13r, TNA; Chandos to Mr. Plunkett, October 29, 1721, Stowe 57, vol. 19, folio 344 r–v, 350r–51v, HL; Chandos to James Phipps, November 13, 1721, Correspondence of the Duke of Chandos, C113/279, folio 35r, TNA; Chandos to Phipps, September 15, 1721, Correspondence of the Duke of Chandos, C113/279, folio 27r, TNA; Chandos to Ambrose Baldwyn, May 11, 1722, Stowe 57, vol. 18, folio 435r–v, HL; Chandos to Mr. Dodson, December 5, 1722, Stowe 57, vol. 22, folios 133r–34v, HL; Dodson, Tinker, and Rice, March 5, 1723, Abstracts of letters received from Coast of Africa, T70/7, folio 40v, TNA; Chandos to Richard Hull, September 24, 1723, Stowe 57, vol. 22, folio 345r–v, HL; Chandos to Capt. General Tinker, December 9, 1723, Stowe 57, vol. 23, folio 135r–v, HL; and Chandos to Captain General John Tinker, December 21 and 29, 1723, Stowe 57, vol. 23, folio 176r–v, HL.

16. Chandos to James Phipps, November 13, 1721, Correspondence of the Duke of Chandos, C113/279, folio 35r, TNA.

17. Royal African Company, Letter Books: Letters Sent, Africa, T70/51, folios 30r–v, 31r–v, 49r–v, TNA.

18. Chandos to James Phipps, December 10, 1720, and December 20, 1720, Stowe 57, vol. 17, folios 298r–v, 305r–v, HL.

19. Chandos to Mr. Phipps, December 19, 1720, Correspondence of the Duke of Chandos, C113/279, folio 5r–v, TNA.

20. Chandos to Ambrose Baldwyn, January 16, 1721, Stowe 57, vol. 18, folio 225r–v ("Lodg'd"), HL; and Chandos to James Phipps, July 21, 1721, Stowe 57, vol. 19, folios 130r–131v ("whereby one"), HL. See also Chandos to James Phipps, January 6, 1721, Stowe 57, vol. 17, folios 314r–316v, HL; and Chandos to Mr. Dodson, December 5, 1722, Stowe 57, vol. 22, folios 132r–133v, HL.

21. Chandos to Ambrose Baldwyn, March 3, 1722, Stowe 57, vol. 18, folio 388r–v, HL.

22. Chandos to James Phipps, August 4, 1721, Stowe 57, vol. 18, folio 256r–v (dragon's blood), HL; Chandos to Mr. Plunkett, October 29, 1721, Stowe 57, vol. 19, folio 350r–v (cardamom), HL; Chandos to Ambrose Baldwyn, March 3, 1722, Stowe 57, vol. 18, folio 387r–v (dragon's blood and cinchona), HL; Chandos to Mr. Phipps, July 27, 1721, Stowe 57, vol. 19, folio 138r–v (ginger), HL; Chandos to Mr. Glynn, January 12, 1723, Stowe 57, vol. 22, folio 162r–163v (bezoar), HL; and Chandos to Captain General John Tinker, December 21 and 29, 1723, Stowe 57, vol. 23, folio 176r–v (rhubarb), 177r–v (ipecacuanha), HL.

23. Chandos to James Phipps, August 4, 1721, Stowe 57, vol. 18, folios 256r–257v, HL.

24. Barcia, *Yellow Demon of Fever*; Breen, *Age of Intoxication*; Cagle, *Assembling the Tropics*; Dorner, *Merchants and Medicines*; Gómez, *Experiential Caribbean*; and Walker, "Medicines Trade."

25. Stewart, "The Edge of Utility," 60–70; and Chandos to Ambrose Baldwyn, March 3, 1722, Stowe 57, vol. 18, folio 389r–v (quotation), 389r–390v, HL.

26. Breen, *Age of Intoxication*; Crawford and Gabriel, "Introduction"; Dorner, *Merchants and Medicines*; Walker, "Medicines Trade"; and Wallis, "Exotic Drugs and English Medicine."

27. Chandos to Ambrose Baldwyn, March 3, 1722, Stowe 57, vol. 18, folio 389r–v, HL.

28. Chandos to Mr. Plunkett, October 29, 1721, Stowe 57, vol. 19, folio 354r–v, HL.

29. Chandos to Ambrose Baldwyn, February 24, 1723, Stowe 57, vol. 22, folio 186r–v, HL.

30. Chandos to Ambrose Baldwyn, August 1, 1723, Stowe 57, vol. 22, folio 293r–v, HL.

31. Chandos to James Phipps, October 5, 1721, Stowe 57, vol. 19, folios 220r–222v, HL.

32. Captain General John Tinker, December 21 and 29, 1723, Stowe 57, vol. 23, folios 179r–180v, HL; and Chandos to Captain General John Tinker, January 7, 1725, Stowe 57, vol. 25, folio 123r–v, HL.

33. Chandos to James Phipps, September 11, 1721, Correspondence of the Duke of Chandos, C113/279, folio 24r, TNA.

34. Chandos to Hans Sloane, December 4, 1721, Sloane 4046, folios 152v–153r, BL.

35. Chandos to Capt. Bonyman, July 7, 1724, Stowe 57, vol. 24, folio 131r–v, HL.

36. For potash, see September 13, 1698, Royal African Company, Letter Books: Letters Sent, Africa, T70/51, folio 10r–v, TNA. For plantations on the Gold Coast, see Law, "'There's nothing grows,'" 122–25, 130–37; and Law, "King Agaja of Dahomey," 154–58. For efforts to establish indigo plantations before 1698, see Kriger, "'Our indico designe.'"

37. Chandos to James Phipps, July 21, 1721, Correspondence of the Duke of Chandos, C113/279, folio 12v, TNA; Chandos to Phipps, December 4, 1721, Correspondence of the Duke of Chandos, C113/279, folio 38r, TNA; Chandos to Mr. Baldwyn, May 11, 1722, Stowe 57, vol. 18, folios 434r–435v, HL; and Chandos to Mr. Hull, September 24, 1723, Stowe 57, vol. 22, folio 344r–v, HL.

38. Chandos to James Phipps, September 11, 1721, Correspondence of the Duke of Chandos, C113/279, folios 24v–25r, TNA.

39. July 2, 1722, Letters from Africa to the Directors of the Royal African Company, C113/274, folios 209v–210r, TNA.

40. Plunket and Archbald, June 30, 1722, Abstracts of letters received from Coast of Africa, T70/7, folio 28r, TNA; Baldwyn, January 25, 1722, Abstracts of letters received from Coast of Africa, T70/7, folio 30r, TNA; and Tinker and Humfreys, May 10, 1724, Abstracts of letters received from Coast of Africa, T70/7, folio 60r, TNA.

41. Richard Hull to Royal African Company, July 2, 1724, Abstracts of letters received from Coast of Africa, T70/7, folios 61v–62r, TNA.

42. Plunkett and Archbold to Royal African Company, June 30, 1722, Abstracts of letters received from Coast of Africa, T70/7, folio 28r, TNA.

43. Chandos to Hans Sloane, December 4, 1721, Sloane 4046, folios 152v–153r, BL.

44. Chandos to Hans Sloane, December 4, 1721, Sloane 4046, folio 152v, BL.

45. Chandos to Hans Sloane, March 23, 1722, Sloane MS 4046, folios 218r–219v, 219r–v ("virtues of which"), BL; Chandos to Ambrose Baldwyn, May 11, 1722, Stowe 57, vol. 18, folio 434r–v ("good Success"), folio 435r–v ("no Sufficient experiment"), HL; and Chandos to Ambrose Baldwyn, July 7, 1722, Stowe 57, vol. 21, folios 13r–14v, HL.

46. Sloane, *Vegetable and Vegetable Substances*, vol. 3, 956–60, 957 (quotations), NHM. See also Pickering, "Putting Nature in a Box," 261–63. Some of the plants were also used as food, adornment, and perfumes.

47. Sloane, *Vegetable and Vegetable Substances*, vol. 3, 916, NHM.

48. Sloane, *Vegetable and Vegetable Substances*, vol. 3, 957–58, NHM. See also Pickering, "Putting Nature in a Box," 255.

49. Sloane, *Vegetable and Vegetable Substances*, vol. 3, 957–58, NHM; and Chandos to Capt. General Tinker, December 9, 1723, Stowe 57, vol. 23, folio 135r–v, HL.

50. Gómez, *Experiential Caribbean*, esp. 129. See also Breen, *Age of Intoxication*; Cagle, *Assembling the Tropics*; Dorner, *Merchants and Medicines*; Schiebinger, *Secret Cures of Slaves*; and Sweet, *Domingos Álvares*.

51. Chandos to Phipps, October 5, 1721, Correspondence of the Duke of Chandos, C113/279, folio 30r, TNA.

52. Cook, "Markets and Cultures," 130–37; and Dorner, *Merchants of Medicines*, 71–76.

53. Dorner, *Merchants of Medicines*, esp. 71–106.

54. Quoted in Cook, "Markets and Cultures," 131.

55. Cook, "Markets and Cultures," 130–33; and Murphy, "Translating the Vernacular."

56. Murphy, "Translating the Vernacular."

57. Breen, "Flip Side"; and Osseo-Asare, *Bitter Roots*.

58. Chandos to James Phipps, January 6, 1721, Stowe 57, vol. 17, folio 317r–v ("strickly examin'd," "situation, Rivers," "by what means"), HL. See also Chandos to Ambrose Baldwyn, January 16, 1721, Stowe 57, vol. 18, folios 225r–226v, HL; and Chandos to James Phipps, October 14, 1721, Stowe 57, vol. 18, folio 318r–v, HL. If linguistic barriers prevented such interviews, Chandos recommended that agents select two or three "of the most likely ones to make House slaves." He predicted that after a year enslaved at the company's factory, agents would be able to communicate with them.

59. Chandos to James Phipps, January 6, 1721, Stowe 57, vol. 17, folio 317r–v, HL. The Royal African Company's plan to interview enslaved Africans as a means of collecting natural knowledge parallels James MacQueen's reliance on enslaved African informants in his geographical study of the Niger River. MacQueen never visited Africa. Instead, he relied

on enslaved informants on the Grenada plantation where he was employed as an overseer. Lambert, *Mastering the Niger*, esp. 88–118.

60. For example, see Chandos to Ambrose Baldwyn, January 16, 1721, Stowe 57, vol. 18, folio 226r–v, HL.

61. Chandos to Hans Sloane, December 7, 1721, Sloane 4046, folio 156r–v, BL.

62. Court of Assistants to Henry Glynn, William Ramsay, and Robert Richardson, January 23, 1722, Letterbooks: Letters Sent to Africa, T70/53, folios 153r–156v, TNA.

63. Chandos to Hans Sloane, December 7, 1721, Sloane 4046, folio 156r–v, BL; and Francis Lynn to Hans Sloane, December 29, 1721, Sloane 4046, folio 166r–v, BL. See also Chandos to James Phipps, November 13, 1721, Correspondence of the Duke of Chandos, C113/279, folio 35r–v, TNA.

64. Chandos to James Phipps, July 21, 1721, Stowe 57, vol. 19, folios 127r–128v, HL. See also Chandos to James Phipps, July 27, 1721, Stowe 57, vol. 19, folio 138r–v, HL; and Chandos to Francis Lynn, July 7, 1722, Stowe 57, vol. 21, folio 10r–v, HL.

65. Chandos to Ambrose Baldwyn, January 16, 1721, Stowe 57, vol. 18, folio 224r–v, HL; and Chandos to Captain General John Tinker, December 9, 1723, Stowe 57, vol. 23, folio 131r–v, HL. Barlow appears to have commanded nine total slaving voyages. For the three Royal African Company voyages, see *Slave Voyages: Trans-Atlantic* (voyage ID 76348, 76536, 76695).

66. Chandos to Mr. Plunkett, October 29, 1721, Stowe 57, vol. 19, folio 345r–v, HL; Chandos to James Phipps, August 4, 1721, Stowe 57, vol. 18, folio 257r–v, HL; Chandos to Ambrose Baldwyn, July 7, 1722, Stowe 57, vol. 21, folio 13r–v, HL; and Chandos to Mr. Dodson, December 5, 1722, Stowe 57, vol. 22, folio 133r–v, HL.

67. Chandos to Captain General John Tinker, December 9, 1723, Stowe 57, vol. 23, folio 140r–v ("grows in great plenty"), HL; and Carney and Rosomoff, *In the Shadow of Slavery*, 3. Although Chandos referred to the *assintee* plant as bitter grass, it is possible that he was describing the plant commonly known as bitter leaf (*Vernonia amygdalina*). Bitter leaf is widely grown and used in both West Africa and Jamaica as a culinary and medicinal plant. See Carney and Rosomoff, *In the Shadow of Slavery*, 179, 235; and Martin and Ruberté, *Edible Leaves*, 65.

68. Chandos to Mr. Glynn, January 12, 1723, Stowe 57, vol. 22, folio 162r–v ("very wholesome"), HL; and Chandos to Dr. Stewart, December 5, 1723, Stowe 57, vol. 23, folio 105r–v ("'tis very probable"), HL.

69. Chandos to Dr. Stewart, December 5, 1723, Stowe 57, vol. 23, folio 105r–v, HL.

70. Pickering, "Putting Nature in a Box," esp. 71, 255. Pickering indicates that the catalog provides geographic information about 5,303 specimens out of the 12,523 total. The 188 African specimens therefore represent 3.5 percent of specimens for which geographic information survives.

71. Mitchell, "'Legitimate commerce,'" 548–53, 564–77.

CHAPTER THREE

1. South Sea Company, "Official copies of letters and instructions from the Court of Directors of the South Sea Company to their Factors abroad," Add. MS 25563, folios 50v–51r, BL.

2. Gould, "Entangled Histories"; Koot, *Empire at the Periphery*, esp. 1–4; and Finucane, *Temptations of Trade*.

3. See, for example, Cromwell, *Smugglers' World*; Norton, *Sacred Gifts*; Prado, *Edge of Empire*; and Warsh, *American Baroque*.

4. Kellman, "Nature, Networks, and Expert Testimony," 383–86; Greenfield, *A Perfect Red*; and Baskes, *Indians, Merchants, and Markets*, 9–11. Cochineal is made from the wingless females of the insect *Dactylopius coccus*.

5. Cañizares-Esguerra, Introduction to *Science in the Spanish and Portuguese Empires*, 2–3; Finucane, *Temptations of Trade*; and Pratt, *Imperial Eyes*, 16.

6. "Account of Books"; "Inquires for Suratte"; and Allen, "The Royal Society and Latin America."

7. Borucki, Eltis, and Wheat, "Slave Trade to Spanish America," esp. 440, 454; Finucane, *The Temptations of Trade*, esp. 12–13; and O'Malley, *Final Passages*, 143–45. Borucki, Eltis, and Wheat estimate that only 300 enslaved Africans out of an estimated total of 56,800 were transported under the Spanish flag between 1701 and 1760.

8. Melville, *South Sea Bubble*, 14; Aiton, "The Asiento Treaty," 167–77; Donnan, "South Sea Company," 419–50; Finucane, *The Temptations of Trade*; Palmer, *Human Cargoes*; Zahedieh, "Merchants of Port Royal," 589–91; and "MINUTES of the Court of Directors of the Governor and Company of Merchants of Great Britain Trading to the South Seas," October 28, 1713, South Sea Company Papers, vol. 1, Add. MS 25495, folios 189r–190v, BL.

9. Brown, "South Sea," 662–78; Klooster, "Inter-Imperial Smuggling," 165–66, 169; Nelson, "Contraband Trade"; O'Malley, *Final Passages*, 163–66, 169–70, 219–63; Palmer, *Human Cargoes*, 9–11; and Walker, *Spanish Politics and Imperial Trade*, 68–72.

10. Henry St. John, Viscount Bolingbroke to the Lords of the Admiralty, July 17, 1713, "Official copys of the most material Letters received by the South Sea Company and other papers," MS Add 25562, folio 3r–v, BL; South Sea Company, "Official copies of memorials, applications, petitions, presentations, addresses, and letters from the Court of Directors of the South Sea Company," Add MS 25559, folios 39r, 45v, BL; and Dewhurst, *Thomas Dover's Life*, xviii–xix. The unauthorized passengers included Dover's secretaries, maids, and (possibly) mistresses as well as Toller's wife and child. Dewhurst, *Thomas Dover's Life*, xix; Dewhurst and Doublet, "Thomas Dover," 112–13; and William Toller, "History of a Voyage to the River of Plate & Buenos Ayres from England" (1715), Mss 3039, Biblioteca Nacional de España, Madrid, Spain.

11. Toller, "History of a Voyage," folio 11v.

12. Toller, "History of a Voyage," folios 15r, 17r–v ("good & Fat"), 17v.

13. Toller, "History of a Voyage," folio 3v.

14. Toller, "History of a Voyage," folios 17v–18v, 18v ("prospect of this Place"), 17v ("not Luxuriant").

15. Finucane, *Temptations of Trade*, 33–34, 42, 60–61, 133.

16. Toller, "History of a Voyage," folios 3v–33v, 21r ("Bay of Castillos"), 29r ("Warwick Bay"). For sailing instructions, see "Directions for Sailing up Rio de La Plata" in Toller, "History of a Voyage," folio 16r.

17. Toller, "History of a Voyage," folios 22v–33v, 22v ("preliminarys"); and Dewhurst and Doublet, "Thomas Dover," 113–14. Six weeks later the crew and passengers of the *Warwick* were still waiting for a Spanish pilot. Dover and the remaining factors decided they could not wait any longer. With the loan of two of the *Warwick*'s boats and the crew to man them, the factors completed the last 140 miles of their journey to Buenos Aires. Once there, Dover hired local pilots to return downriver and guide the royal naval vessel into harbor. The *Warwick* finally anchored at Buenos Aires in early September.

18. Toller, "History of a Voyage," folio 20v.

19. Brown, "South Sea Company," 670; *Slave Voyages: Trans-Atlantic* (voyage ID 76318); "List of plants received from John Burnett," Sloane 4072, folio 295r, BL; and "MINUTES of the Court of Directors of the Governor and Company of Merchants of Great Britain Trading to the South Seas and other Parts of America and For Encouraging the Fishery," April 25, 1716, South Sea Company Papers, vol. 3, Add MS 25496, folio 225r–v, BL. For more about Burnet's collecting on the *Wiltshire*, see this book's introduction. For the impact of the War of the Quadruple Alliance on the *asiento* trade, see Finucane, *Temptations of Trade*, 57–63.

20. Finucane, *Temptations of Trade*, 12, 26–27; and John Burnet to Hans Sloane, October 6, 1725, Sloane 4048, folio 70r, BL.

21. Burnet's tendency to refer collectively (for example, "some natural productions") to the objects he sent makes it impossible to fully quantify the extent of his collection. But his correspondence and the catalog to Sloane's museum indicate eighty-five discrete specimens. Groups of specimens referred to in the sources collectively, such as "some herbs" or a box of shells, suggest that the actual count easily exceeded those that can be individually identified. For more on Burnet's collecting, see Murphy, "A Slaving Surgeon's Collection."

22. John Burnet to James Petiver, December 26, 1716, Sloane MS 3322, folio 97r–v, BL; John Burnet to Hans Sloane, April 6, 1722, Sloane 4046, folio 227r, BL; Hans Sloane, *Vegetable and Vegetable Substances*, vol. 3, folio 797r–v; and "List of plants received from John Burnett," Sloane 4072, folio 295r, BL.

23. Hans Sloane, Catalogue of Minerals, vol. 1, 160, Paleontology Library, MSS SLO, NHM; Hans Sloane, Catalogue of Minerals, vol. 2, 2, Paleontology Library, MSS SLO, NHM; and Hans Sloane, Catalogue of Minerals, vol. 3a, [unpaginated], Paleontology Library, MSS SLO, NHM. For "green gold," see Schiebinger, *Plants and Empire*, 7.

24. John Burnet to Hans Sloane, April 6, 1722, Sloane 4046, folio 227v, BL; John Burnet to Hans Sloane, August 6, 1723, Sloane 4047, folio 29r–v, BL; and Grahn, "Catagena and Its Hinterland." For Cartagena's Black healers in the long seventeenth century, see Gómez, *The Experiential Caribbean*.

25. John Burnet to James Petiver, December 26, 1716, Sloane MS 3322, folio 97r–v (quotation), BL; Grahn, "Catagena and Its Hinterland," 180–81; and Parrish, *American Curiosity*, 215–306. Burnet's collections included at least one object made by a Native American, "the head of an Indian dart from Carthagena in America. It is made of the Beryll or darkish coloured Chrystall." Hans Sloane, Miscellanies Catalogue, item 1344, Centre of Anthropology, Department of Africa, Oceania, and the Americas, British Museum. For the medical marketplaces of early modern Cartagena see Gómez, *Experiential Caribbean*.

26. John Burnet to Hans Sloane, March 17, 1725, Sloane 4047, folios 329r–330r, 329v ("personal love"), BL; South Sea Company, "Official copies of letters and instructions from the Court of Directors of the South Sea Company to their Factors abroad," November 19, 1724, Add. MS 25564, folio 79r ("the Esteem" and "Signal Service"), BL; John Burnet to Hans Sloane, April 6, 1722, Sloane 4046, folios 227r–228v, 228r ("Padre Guardian"), BL; and John Burnet to Hans Sloane, August 6, 1723, Sloane 4047, folio 29r–v, 29v ("Moss" and "Lapis Butleri"), BL.

27. John Burnet to Hans Sloane, September 1722, Sloane 4046, folios 287r–288v, BL. See also John Burnet to Hans Sloane, April 6, 1722, Sloane 4046, folios 227r–228v, BL; John Burnet to Hans Sloane, August 6, 1723, Sloane 4047, folios 29r–30v, BL; and Halley, "The Longitude of Carthagena," 237–38.

28. John Burnet to Hans Sloane, July 17, 1725, Sloane 4048, folio 26r–v, 26v ("acceptable"), BL. See also John Burnet to Hans Sloane, August 6, 1723, Sloane 4047, folios 29r–30v, BL; John Burnet to Hans Sloane, February 24, 1724, Sloane 4047, folios 323v–324r, BL; John Burnet to Hans Sloane, April 2, [1724?], Sloane 4047, folio 164r, BL; John Burnet to Hans Sloane, July 17, 1724, Sloane MS 4047, folio 198r, BL; John Burnet to Hans Sloane, March 17, 1725, Sloane 4047, folios 329r–330r, BL; John Burnet to Hans Sloane, April 7, 1725, Sloane MS 4047, folio 333r, BL; John Burnet to Hans Sloane, January 5, 1726, Sloane 4048, folio 120r–v, BL; and Halley, "The Longitude of Carthagena," 237–38.

29. John Burnet to Hans Sloane, July 17, 1724, Sloane MS 4047, folio 198r ("new books"), BL; and John Burnet to Hans Sloane, April 2, 1724 [?], Sloane 4047, folio 164r, BL.

30. John Burnet to Hans Sloane, January 5, 1726, Sloane 4048, folio 120r ("good graces"), BL; and John Burnet to Hans Sloane, April 6, 1722, Sloane 4046, folios 227r–228v, 228r ("would use"), 228v ("diligent discharge"), BL. For other requests for a promotion or a raise, see John Burnet to Hans Sloane, November 30, 1716, Sloane 4044, folio 250r, BL; "Official copies of letters and instructions from the Court of Directors of the South Sea Company to their Factors abroad," April 30, 1718, South Sea Company Papers, Add. MS 25563, folio 164r, BL; John Burnet to Hans Sloane, August 6, 1723, Sloane 4047, folio 29v, BL; John Burnet to Hans Sloane, February 24, 1724, Sloane 4047, folios 323v–324r, BL; and John Burnet to Hans Sloane, March 17, 1725, Sloane 4047, folios 329r–330r, BL.

31. John Burnet to Hans Sloane, October 10, 1727, Sloane MS 4049, folio 50v, BL.

32. Brown, "South Sea Company," 662–78; John Burnet to Hans Sloane, October 10, 1736, Sloane 4054, folios 314r–315v, BL; July 8, 1736, Journal Books of Scientific Meetings, Collections from the Royal Society, 1660–1800, vol. 15, p. 368.

33. John Burnet to James Petiver, April 16, 1718, Sloane 4065, folio 285r, BL.

34. Parsons and Murphy, "Ecosystems under Sail," 527–28.

35. *Slave Voyages: Trans-Atlantic* (voyage ID 75696); "Official copies of letters and instructions from the Court of Directors of the South Sea Company to their Factors abroad," April 30, 1718, South Sea Company Papers, Add. MS 25563, folio 158r, BL; and O'Malley, *Final Passages*, 229. Historian Colin Palmer's analysis of the South Sea Company's trade indicates that the linked factories at Portobello and Panama had one of their busiest years in 1718. In 1718, 9.9 percent of slaves arriving in Portobello were ill, compared with an average of 5.7 percent over the period 1716–25. See table 12, Slaves to Panama and Porto Bello, 1715–38; and table 23, Health of Slaves Arriving at Porto Bello, 1716–25, in Palmer, *Human Cargoes*, 103, 118.

36. John Burnet to James Petiver, April 16, 1718, Sloane 4065, folio 287r, BL.

37. John Burnet to James Petiver, April 16, 1718, Sloane 4065, folio 286v, BL.

38. Thomas Knapp to Hans Sloane, October 23, 1718, Sloane 4045, folio 159r, BL.

39. Quoted in Stearns, "James Petiver," 342.

40. Petiver, *Hortus Peruvianus Medicinalis*, 1, tab. 2; and James Petiver to Thomas Dover, March 31, 1715, Sloane 3340, folios 261r–262v, BL.

41. "An Account of a Book entituled, *Gazophylacii Naturae & Artis*," 342–52, 350 (quotation); and William Sherard to Richard Richardson, November 22, 1722, *Ms. Radcliffe Trust C.4*, folio 86r–v, Bodleian Library, quoted in Stearns, *Science in the British*, 385n156.

42. James Petiver to George Jesson, July 20, 1716, Sloane MS 3340, folio 252r–v, BL; and Hans Sloane, *Vegetable and Vegetable Substances*, vol. 2, folio 805r–v, NHM.

43. James Petiver to David Patton, July 2, 1715, Sloane 3340, folios 172v–173r, BL; James Petiver to David Patton, July 8, 1717, Sloane 3340, folios 330v–331r, BL; James Petiver to

William Toller, November 19, 1716, Sloane 3340, folios 275v–276v, BL; and Murphy, "Collecting Slave Traders," 662.

44. Thomas Dover to [James Petiver], March 29, 1716, Sloane MS 3322, folio 91r, BL. Whereas most South Sea Company factors had a mercantile background, Dover had previously been a physician, a privateer, and involved in the slave trade as a surgeon, captain, and ship owner. Dewhurst, *Thomas Dover's Life*, xi–xvii; Dewhurst and Doublet, "Thomas Dover," 107–11; and Strong, *Dr. Quicksilver*, 67–132.

45. Thomas Dover to [James Petiver], March 29, 1716, Sloane MS 3322, folio 91r, BL.

46. Francis Hall to William Sherard, July 8, 1727, Sherard Papers, MS 255, folio 595r–v, RS.

47. John Burnet to Hans Sloane, April 6, 1722, Sloane 4046, folio 227v ("skin stuffed"), BL; John Burnet to Hans Sloane, September 1722, Sloane 4046, folios 287r–288v, BL; and John Burnet to Hans Sloane, August 6, 1723, Sloane 4047, folio 29r–v, BL.

48. John Burnet to Hans Sloane, April 6, 1722, Sloane 4046, folios 227r–228v, BL; and John Burnet to Hans Sloane, August 6, 1723, Sloane 4047, folio 29v, BL.

49. O'Malley, *Final Passages*, 221–24, 233–43.

50. Houstoun observed 294 of the 671 plants he described in his "Catalogus Plantarum" while in Jamaica. For more on Houstoun, see chapter 4. William Houstoun, "Catalogus Plantarum in America observatarum," ca. 1730–33, MSS Banks Coll Hou, NHM.

51. "OFFICIAL copies of letters and instructions from the Court of Directors of the South Sea Company to their Factors abroad, persons in their employ, and various public companies and officials," 7 vols., Add. MS 25563, vol. II, folio 44r ("Person of Skill," 44r), folios 60v–61r, ("assist our Factory," 60v), BL.

52. James Petiver to William Toller, November 19, 1716, Sloane MS 3340, folios 275v–276v, BL.

53. Francis Hall to William Sherard, July 8, 1727, Sherard Papers, MS 255, folio 595r–v, 595r ("Tiggers skin" and "small piece"), 595v ("Skin of an Amphibious Bird"), RS. See also Francis Hall to William Sherard, March 30, 1726, Sherard Papers, MS 255, folio 598r–v, RS. Sloane's museum also included specimens gathered by Hall. See Sloane, *Vegetable and Vegetable Substances*, vol. 3, folio 1031r–v, items 9367–70, NHM.

54. Clokie, *Account of the Herbaria*, 80; Burkart, "*Hortus Elthamensis* de Dillenius," 375–77; Fielding-Druce Herbarium, Oxford University Herbaria; and Oxford University Herbaria On-Line Database, https://herbaria.plants.ox.ac.uk/bol/oxford/Explore.

55. James Douglas, "A short account of the different kinds of Ipecacuanha," MS Hunter D422, UGSC; and James Douglas, "The Description and Natural History of the Animal called Armadillo or the hog in armour from South America or the little American hog in Armour, by J. D.," MS Hunter D516, UGSC.

56. The surviving archival evidence can make it difficult to establish a timeline of William Houstoun's movements. A recent biographical sketch of Houstoun suggests that he did not arrive in the New World until late 1730 and that he graduated with his medical degree from Leiden in 1729 after studying with Herman Boerhaave. However, Philip Miller makes multiple references to receiving American plants from Houstoun in 1728 and 1729. Moreover, the South Sea Company's records note that "Mr. William Houstoun" was "especially appointed by the Company, Surgeon of their Sloop Assiento," under the command of Captain Uring by October 1729. One source of confusion is that William Houstoun's kinsman James Houstoun was also employed by the South Sea Company during the same period. James Houstoun stated in his memoirs that he graduated from Leiden after studying with Boerhaave, perhaps explaining the source of some of this confusion. See D.W. [Daniel Westcomb] to [Edward Pratter],

October 30, 1729, "Official copies of letters and instructions from the Court of Directors of the South Sea Company to their Factors abroad," Add. MS 25566, folio 141r, BL; Miller, *Gardeners Dictionary*; Houstoun, *Works of James Houstoun*, 50; and Rose, "William Houstoun."

57. William Houstoun, Botanical manuscripts and drawings of plants collected in Central America, Jamaica and Cuba, ca. 1730–33, MSS Banks Coll Hou, NHM; and William Houstoun to Hans Sloane, December 9, 1730. Sloane 4051, folio 141r–v, BL.

58. William Houstoun to Hans Sloane, December 9, 1730. Sloane 4051, folio 141r–v, BL.

59. Hans Sloane, *Vegetable and Vegetable Substances*, vol. 3, 9115, 9116, 9229, NHM.

60. Pickering, "Putting Nature in a Box," 104.

61. Miller, *Gardeners Dictionary*; and Stearn, "Philip Miller," 296.

62. Miller, *Figures of the most Beautiful*, 30; Miller, *Gardeners Dictionary*; and Journal Books of Scientific Meetings, Collections from the Royal Society, 1660–1800, vol. 21, March 29, 1753, vol. 21, 300.

63. Hevly, "Nomenclatural History," 529.

64. John Martyn, *Historia plantarum rariorum*, 25, 42; Hevly, "Nomenclatural History," 527–29; Rand, *Horti Medici Chelseiani*; and Ewan, "Plant Collectors in America," 30.

65. Banks published a small fraction of the plants described and illustrated in Houstoun's manuscripts under the title *Reliquae Houstounianae* (1781). *Reliquae Houstounianae* featured just twenty-five species out of the hundreds featured in Houstoun's papers. Dandy, *Sloane Herbarium*, 139; and Houstoun, *Reliquae Houstounianae*.

66. Ewan, "Plant Collectors in America," 32, 44–47.

67. Klooster, "Inter-Imperial Smuggling," 141.

CHAPTER FOUR

1. William Houstoun to Hans Sloane, March 4, 1732, Sloane 4052, folio 82r–v, BL. See also Candler, *Minutes of the Common Council*, October 3, 1732, 2:5.

2. O'Malley, *Final Passages*, 240 (quotation), 162–64, 239–41.

3. Historians Alex Borucki, David Eltis, and David Wheat noted that the slave trade to Spanish territories was the only branch of the transatlantic slave trade "that delivered the majority of its captives outside the limits of the law and official policy." Borucki, Eltis, and Wheat, "Slave Trade to Spanish America," 454.

4. Houstoun, "Account of the Contrayerva," 195–98; Journal Books of Scientific Meetings, Collections from the Royal Society, 1660–1800, vol. 14, 12 (November 4, 1731), 136 (May 11, 1732), 168 (October 26, 1732), and 223 (January 18, 1733).

5. O'Malley, *Final Passages*, 226, 236; and Palmer, *Human Cargoes*, 59–64.

6. William Houstoun to Hans Sloane, December 9, 1730, Sloane Manuscripts 4051, folio 142r, BL; and "Official copies of letters and instructions from the Court of Directors of the South Sea Company to their Factors abroad," October 30, 1729, South Sea Company Papers, Add. MS 25566, folio 141r, BL.

7. For an important exception to this trend, see Cook, *Matters of Exchange*.

8. Four voyages of the *Assiento* (also spelled *Asiento*) that date to Houstoun's tenure as a ship surgeon under the command of Captain Uring appear in the Intra-American Slave Trade database. The average number of captives disembarked from these voyages is ninety-four. See *Slave Voyages: Intra-American*.

9. William Houstoun to Hans Sloane, December 9, 1730. Sloane 4051, folios 141r–142r, 141v ("great many Plants"), BL; and William Houstoun, Botanical manuscripts and drawings of

plants collected in Central America, Jamaica and Cuba, ca. 1730–33, MSS Banks Coll Hou, NHM. The first and longest section of the four parts within Houstoun's botanical manuscript includes 671 different plants.

10. Carroll, *Blacks in Colonial Veracruz*, 3.

11. Carroll, *Blacks in Colonial Veracruz*, 14–19, 73–81; Borucki, Eltis, and Wheat, "Slave Trade to Spanish America," 458, 466; Palmer, *Human Cargoes*, 102–8; and O'Malley, *Final Passages*, 163–64, 238–41.

12. Palmer estimates 19,662 captive Africans were sold by the South Sea Company in the linked factories of Portobello and Panama, compared to just 3,011 for Veracruz. Palmer, *Human Cargoes*, 103, 108.

13. *Slave Voyages: Intra-American* (Voyage ID 110071, 110072); and "An account of the Contrayerva by William Houstoun Surgeon in the Service of the hon'ble South Sea Company," Record Book Originals, RBO/16/20, folio 139r-v, RS. Sloane's museum included two objects credited to a Captain Uring, which was also the name of Houstoun's commanding officer on the *Assiento* (or *Asiento*): a bird described as a cockatoo from Darien (Panama) and pods of the Acacia tree from Veracruz. Both entries also referenced Houstoun, thus raising the strong possibility that the two objects were given to Sloane by the *Assiento*'s captain. Hans Sloane, *Vegetable and Vegetable Substances*, vol. 3, folio 1024r-v, NHM; and Hans Sloane, *Fossils*, vol. 5, folio 201r-v, NHM.

14. Catesby, *The Natural History of Carolina*, 1:viii, 50; and James Petiver, "Directions for George to take with him," October 18, 1698, Sloane 3333, folio 235r-v, BL.

15. Safier, *Measuring the New World*, 9.

16. *Slave Voyages: Intra-American* (Voyage ID 110071); Miller, *Botanicum officinale*, 147–48, 148 ("very good against"); William Houstoun to Hans Sloane, December 9, 1730, Sloane 4051, folio 141r-v, BL; and Houstoun, "Account of the Contrayerva," 195–98.

17. William Houstoun to Hans Sloane, December 9, 1730, Sloane 4051, folio 141r-v, BL; and Houstoun, "Account of the Contrayerva," 195–98. For Thiéry, see Schiebinger, *Plants and Empire*, 42.

18. William Houstoun to Hans Sloane, December 9, 1730, Sloane 4051, folio 141r-v, BL; Houstoun, "An Account of the Contrayerva," Record Book Originals, RBO/16/20, folios 137r–140v, RS; November 4, 1731, Journal Books of Scientific Meetings, Collections from the Royal Society, 1660–1800, vol. 14, folios 12r–13v; and Miller, *The Gardeners Dictionary*. Houstoun's description of contrayerva was cited in his nomination to become a member of the Royal Society in 1732. See October 26, 1732, Journal Books of Scientific Meetings, Collections from the Royal Society, 1660–1800, vol. 14, 168.

19. Carroll, *Blacks in Colonial Veracruz*, 4; and Gänger, "World Trade in Medicinal Plants." In the early eighteenth century, Spain imported an average of five tons of jalap root each year, most of which was reexported to the broader European drug market.

20. William Houstoun to Hans Sloane, December 9, 1730, Sloane 4051, folios 141r–142r, 141r (quotations), BL. See also William Houstoun to Hans Sloane, March 5, 1731, Sloane 4052, folio 82r-v, BL; Miller, *The Gardeners Dictionary*, unpaginated; and Romans, *A concise natural history*, 154. According to Bernard Romans, the jalap plants Houstoun introduced to Jamaica were destroyed by hogs during one of his absences from the island.

21. Candler, *Minutes of the Common Council*, October 3, 1732, 2:6. See also "Agreement with Dr. William Houstoun the Botanist," October 4, 1732, CO5/670, folios 1r–2v, TNA. For the central place of Jamaica within the South Sea Company's operations, see O'Malley, *Final Passages*, 219–63, esp. 221–24.

22. T. Mun, *England's Treasure by Forraign Trade* (Oxford, 1667), quoted in Drayton, *Nature's Government*, 57. See also Schiebinger, *Plants and Empire*, 7–22, 73–104.

23. In this literature, "bioprospecting" is often used by individuals who believe that the use of traditional knowledge is a necessary part of the modern pharmacological research process, while "biopiracy" is employed by those who see it as a theft of communal resources. For debates over these practices and the associated terms, see, for example, Isaac and Kerr, "Bioprospecting or Biopiracy"; Merson, "Bio-Prospecting or Bio-Piracy"; Robinson, *Confronting Biopiracy*; and Shiva, "Bioprospecting as Sophisticated Biopiracy."

24. Schiebinger, *Plants and Empire*, 35–44.

25. Candler, *Minutes of the Common Council*, October 3, 1732, 2:5.

26. Delbourgo, *Collecting the World*; and Wulf, *Brother Gardeners*, 34–47, 90–92, 120.

27. Stearns, *Science in the British Colonies*, 328; and Royal Society, Journal Books of Scientific Meetings, Collections from the Royal Society, 1660–1800, vol. 14, 136.

28. Candler, *Minutes of the Common Council*, October 3, 1732, 2:5.

29. All of Houstoun's individual patrons subscribed to at least one copy of Catesby's *Natural History of Carolina*, the first major illustrated natural history of British America. Two of Houstoun's patrons, Sloane and Charles Du Bois, additionally sponsored the collecting trip upon which the text was based. Other supporters of Houstoun's expedition were subscribers to John Bartram's annual boxes of North American seeds. Brigham, "Mark Catesby"; Grieve, *Transatlantic Gardening Friendship*, 17–18, 23; and Wulf, *Brother Gardeners*, 86–87, 99.

30. Houstoun's patrons included six individuals and two groups. These patrons and the annual amount each pledged were Charles Du Bois (£10); Charles Lennox, Duke of Richmond (£30); James Edward Oglethorpe (£5); Robert James Petre, Baron Petre (£50); Sir Hans Sloane (£20); James Stanley, Earl of Derby (£50); Society of Apothecaries (£20); and the Trustees of Georgia (£15). The Society of Apothecaries was Miller's employer at Chelsea Physica Garden. Harman Verelst to Hans Sloane, February 13, 1733, Sloane 4053, folio 164r, BL.

31. Oglethorpe, "Some Account of the Designs," 6. For "collector of drugs," see McPherson, *Diary of the First Earl of Egmont*, February 19, 1734, 3:32. See also "Copy of Mr. Houstoun's Instructions," October 12, 1732, CO5/670, folios 2r–3v, TNA.

32. Stewart, "'Policies of Nature,'" 474–82. See also Coleman, *Colonial Georgia*, 13–24, 111–18; Greene, "Travails of an Infant Colony," 283–85; Sweet, "A Misguided Mistake," 1–7; and Wood, "The Earl of Egmont," 85–86.

33. Stewart, "Public Lectures and Private Patronage"; Stewart, *Rise of Public Science*; Stearns, *Science in the British Colonies*, 326–27; Holland, "Beginning of Public Agricultural Experimentation," 276–77; and Sweet, "A Misguided Mistake," 1–7.

34. Candler, *Minutes of the Common Council*, October 3, 1732, 2:5 ("beg[ged] of the Hon'ble Trustees"); Sedgwick, *House of Commons*, 20–21, 121–22; and Sackler Archive Resource, Royal Society of London, http://royalsociety.org/library/collections/biographical-records/. For Sloane's influence with the South Sea Company, see Stewart, "The Edge of Utility," 60–61.

35. William Houstoun to James Oglethorpe, December 21, 1732, in Coleman and Ready, *Original Papers*, 20:4; Robert Millar to the Trustees of Georgia, December 10, 1734, in Coleman and Ready, *Original Papers*, 20:116; Robert Millar to the Trustees of Georgia, June 20, 1735, in Coleman and Ready, *Original Papers*, 20:396; Robert Millar to Trustees of Georgia, July 7, 1736, in Candler, *Original Papers*, 21:193; and Robert Millar to the Trustees of Georgia, May 26, 1738, in Candler, Northen, and Knight, *Original Papers*, 22:150–51.

36. Finucane, *Temptations of Trade*, 97–108.

37. William Houstoun to James Oglethorpe, December 21, 1732, in Coleman and Ready, *Original Papers*, 20:4. See also Candler, *Minutes of the Common Council*, 2:5; Benjamin Martyn to James Oglethorpe, March 31, 1733, in Coleman and Ready, *Trustees' Letter Book*, 29:9; Hancock, *Oceans of Wine*, 118.

38. William Houstoun to James Oglethorpe, December 21, 1732, in Coleman and Ready, *Original Papers*, 20:4.

39. William Houstoun to James Oglethrope, January 25, 1733, in Coleman and Ready, *Original Papers*, 20:6 (quotations), 5–6.

40. Mr. Cochrane to Philip Miller, September 11, 1733, in Coleman and Ready, *Original Papers*, 20:33–34, 33 ("long and severe"); and Johnston, "Doctor William Houstoun," 12.

41. Candler, *Minutes of the Common Council*, February 20, 1734, and March 6, 1734, 2:58–59, 60–61; McPherson, *Diary of the First Earl of Egmont*, February 19, 1734, 2:32; "Agreement with and Instructions for Robert Millar," CO5/670, folio 1r–2v, TNA; Stearns, *Science in the British Colonies*, 329–30; and Dandy, *The Sloane Herbarium*, 165.

42. Robert Millar to Trustees of Georgia, May 26, 1738, in Candler, Northen, and Knight, *Original Papers*, 22:151.

43. Robert Millar to Hans Sloane, August 8, 1734, Sloane 4034, folio 250r–v, BL; and Robert Millar to the Trustees of Georgia, December 10, 1734, in Coleman and Ready, *Original Papers*, 20:115–16, 116 ("Liberty to go Passenger").

44. Robert Millar to Trustees of Georgia, December 10, 1734, in Coleman and Ready, *Original Papers*, 20:116 (quotations), 116–17.

45. Robert Millar to the Trustees of Georgia, June 20, 1735, in Coleman and Ready, *Original Papers*, 20:395–98; Robert Millar to the Trustees of Georgia, July 7, 1736, in Candler, *Original Papers*, 21:192–93.

46. Robert Millar to Trustees of Georgia, June 20, 1735, in Coleman and Ready, *Original Papers*, 20:395–98; Robert Millar to Trustees of Georgia, July 7, 1736, in Candler, *Original Papers*, 21:192–94, 193 ("many other Trees"); Miller, *Botanicum officinale*, 72–73, 244 ("almost justled"), 73 ("excellent pectoral"). Almost half (nineteen of forty) of the plants Philip Miller credited Robert Millar with introducing into the Chelsea Physic Garden were collected during Millar's time in Cartagena. See Miller, *Gardeners Dictionary*, unpaginated.

47. Robert Millar to the Trustees of Georgia, July 7, 1736, in Candler, *Original Papers*, 21:194–95.

48. For Montijo's election to the Council of the Indies, see South Sea Company, "MINUTES of the Court of Directors of the Governor and Company of Merchants of Great Britain Trading to the South Seas," June 23, 1737, Add. MS 25509, folio 1r, BL. The Duke of Richmond, a privy counsellor and member of the Royal Society, obtained the letter of introduction on Millar's behalf. Robert Millar to Trustees of Georgia, February 12, 1737, in Candler, *Original Papers*, 21:282.

49. Conde de Montijo to Juan Antonio de Vizarrón y Eguiarreta, January 7, 1737, in Candler, *Original Papers*, 21:292–93; Robert Millar to the Trustees of Georgia, July 7, 1736, in Candler, *Original Papers*, 21:191–95; Robert Millar to the Trustees of Georgia, July 22, 1737, in Candler, *Original Papers*, 21:491–92; and Robert Millar to the Trustees of Georgia, May 26, 1738, in Candler, Northen, and Knight, *Original Papers*, 22:150–52. It is unclear whether Montijo himself was the author of this deception or if he merely repeated what he had been told.

50. Robert Millar to the Trustees of Georgia, May 26, 1738, in Candler, Northen, and Knight, *Original Papers*, 22:150–52.

51. Robert Millar to the Trustees of Georgia, May 26, 1738, in Candler, Northen, and Knight, *Original Papers*, 22:151–52.

52. Robert Millar to the Trustees of Georgia, May 26, 1738, in Candler, Northen, and Knight, *Original Papers*, 22:150–52. The *asiento* contract promised South Sea Company resident factors much greater freedom of movement than mariners employed by the company.

53. Roberts, *Manuscripts of The Earl of Egmont*, August 23, 1738, 2:506 ("seen no fruits"). For grapes in colonial Georgia, see Church, *Oglethorpe*, 95–98; and Pinney, *History of Wine*, 25–28. These vines, like all European varietals introduced into North America before the nineteenth century, fell prey to diseases native to the New World. In the nineteenth century, wine producers discovered that grafting European varietals onto American rootstocks alleviated the problem.

54. For Millar sending seeds to Miller, see Robert Millar to Trustees of Georgia, July 7, 1736, in Candler, *Original Papers*, 21:192, 193; and Robert Millar to the Trustees of Georgia, December 10, 1734, in Coleman and Ready, *Original Papers*, 20:117. For Millar in South Carolina, see McPherson, *Manuscripts of The Earl of Egmont*, January 16–23, 1740, 3:100. As late as May 1738, Millar promised the Georgia trustees that he would deliver *Ipecacuanha* and other plants to Georgia, supporting the idea that he may have left these plants with the trustees' agent in South Carolina. Robert Millar to Trustees of Georgia, May 26, 1738, in Candler, Northen, and Knight, *Original Papers*, 22:150–52.

55. For the 212 seeds, see Robert Millar to Hans Sloane, July 22, 1737, Sloane 4055, folio 147r–v, BL; and Robert Millar to Hans Sloane, [1737], Sloane 4059, folio 356r–v, BL. For Sloane's catalog, see Pickering, "Putting Nature in a Box," 104.

56. Miller, *Figures of the most Beautiful*, 1:74.

57. Miller, *Gardeners Dictionary*, unpaginated; Clokie, *An Account of the Herbaria*, 86, 186, 212; and Ewan, "Plant Collectors in America," 44–47.

58. Philip Miller to Richard Richardson, June 18, 1734, Bodleian card catalog, Early Modern Letters Online, http://emlo.bodleian.ox.ac.uk/; and Aiton, *Hortus Kewensis*, x–xi.

59. O'Malley, *Final Passages*, 163.

60. Cook, *Matters of Exchange*; Margócsy, *Commercial Visions*; Schiebinger and Swan, *Colonial Botany*; and Stewart, "Global Pillage."

CHAPTER FIVE

1. Brock, "Dru Drury's *Illustrations of Natural History*"; and Hancock and Douglas, "William Hunter's Goliath Beetle," 218–22.

2. Dru Drury to Capt. Mayle [Male], October 30, 1766, Drury Letter Book, 83; "A Catalogue of the Most Capital Assemblage of Insects Probably Ever Offered to Public Sale; Consisting of upwards of Eleven Thousand different Specimens, collected . . . by Mr. Dru Drury . . ." [1805], Oxford University Museum of Natural History; Brock, "Dru Drury's *Illustrations*," 259–65; Hancock and Douglas, "William Hunter's Goliath Beetle," 222; Smith, "Memoir of Dru Drury"; and Salmon, *Aurelian Legacy*, 114–15.

3. This includes vessels registered in both Britain and British America. See *Slave Voyages: Trans-Atlantic*.

4. French forces recaptured their former territory during the American Revolutionary War. This marked the end of Britain's first colony in West Africa. Brown, "Empire without America," 84–100; Dziennik, "'Till these Experiments be Made'"; Martin, *British West African Settlements*; and Newton, "Naval Power and the Province of Senegambia."

5. Newton, "Slavery, Sea Power and the State," 173. See also Martin, *British West African Settlements*, 8–56.

6. For Drury's belief that the specimen had not traveled far, see Dru Drury to Capt. Mayle [Male], October 24, 1768, Drury Letter Book, 141. For the *Goliathus goliatus*, see Hancock and Douglas, "William Hunter's Goliath Beetle," 224.

7. Dru Drury to Nonus Parke, April 1769, Drury Letter Book, 160. Drury composed but never sent a letter to Parke in which he explained why he sent him copies of the Goliath print. Drury noted in his letter book that Parke's visit allowed him to make the request in person. For Parke's slaving voyages, see *Slave Voyages: Trans-Atlantic* (voyage ID 77918, 90700, 90701, 91573, 91574).

8. Brock, "Dru Drury's *Illustrations*"; Hancock and Douglas, "William Hunter's Goliath Beetle," 218–22; and Dru Drury, "Giving an account of the printing expenses incurred by the publication of Illustrations of Natural History (1770–82)," SB o D.11, NHM. For the infamous exchange between da Costa and Hunter, see Brock, "Dru Drury's *Illustrations*," 260–61.

9. Hancock and Douglas, "William Hunter's Goliath Beetle," 220.

10. The five confirmed recipients of the Goliath print were the slaving captains Richard Cowley, Nonus Parke, and Thomas Williams; the royal naval captain Thomas Male; and Capt. Thomas Miller of West Florida. The business that brought Miller to West Central Africa is unknown, and the voyage does not appear in the Trans-Atlantic Slave Trade Database. It is possible that Miller was engaged in the slave trade as he was in the process of establishing a rice plantation in the newly British territory of West Florida. He previously petitioned for 20,000 acres of land in West Florida and by 1767 owned a rice plantation in the new colony with at least fourteen enslaved Africans. For Drury's letters referencing the Goliath print, see Dru Drury to Richard Cowley, November 20, 1769, Drury Letter Book, 176; Dru Drury to Capt. Mayle [Male], October 24, 1768, Drury Letter Book, 141; Dru Drury to Thomas Miller, [March 1771], Drury Letter Book, 233; Dru Drury to Nonus Parke, February 10, 1771, Drury Letter Book, 228; and Dru Drury to Capt. Williams, December 14, 1768, Drury Letter Book, 147. For Miller, see Thomas Miller to John Ellis, April 16, 1767, vol. 2, Ellis Manuscripts, LLS; Rauschenberg, "John Ellis," 6, 29; and Rea, *Major Robert Farmar*, 44, 129, 166n8.

11. Dru Drury to Capt. Mayle [Male], October 24, 1768, Drury Letter Book, 141; and Dru Drury to Capt. Williams, December 14, 1768, Drury Letter Book, 147.

12. Dru Drury to Capt. Mayle [Male], October 24, 1768, Drury Letter Book, 141.

13. Dru Drury to Capt. Williams, December 14, 1768, Drury Letter Book, 147 ("Enquire if they know"); Dru Drury to Nonus Parke, April 1769, Drury Letter Book, 160; and Dru Drury to Nonus Parke, February 10, 1771, Drury Letter Book, 228 ("But above all things").

14. Dru Drury to Capt. Mayle [Male], October 24, 1768, Drury Letter Book, 141.

15. Drury, *Illustrations of Natural History*, 1:xv. Drury also stressed that his descriptions would indicate to future publishers of his book how each specimen depicted in the prints should be colored, thus ensuring the engravings' accuracy in any subsequent or posthumous editions.

16. Bleichmar, *Visible Empire*, 46 (quotation), 43–77. See also Daston and Galison, *Objectivity*, 55–113.

17. Bleichmar, *Visible Empire*, 43–77; Drury, "Giving an account of the printing expenses incurred by the publication of Illustrations of Natural History (1770–1782)," SB o D.11, NHM; and Dru Drury to Thomas Miller, [March 1771], Drury Letter Book, 233.

18. Margócsy, *Commercial Visions*, 25 (quotation), 29–73.

19. Dru Drury to Capt. Williams, December 14, 1768, Drury Letter Book, 147; and Ledger, Drury Letter Book, folio 1r. For more on the equipment and collecting practices of eighteenth-century entomologists, see Salmon, *Aurelian Legacy*, 55–82.

20. Dru Drury, "Directions for Collecting Insects in Foreign Countries," BL; Noblett, "Dru Drury's 'Directions'"; and Dru Drury to Mr. Woolin, November 13, 1767, Drury Letter Book, 112. According to William Noblett, the British Library's copy of Drury's "Directions" is the only known copy extant. He also notes that it represents the first printed instructions for entomological collecting distributed for free.

21. Dru Drury, "Directions for Collecting Insects in Foreign Countries," BL. See also Noblett, "Dru Drury's 'Directions.'"

22. Dru Drury, "Directions for Collecting Insects in Foreign Countries," BL, 1–3. See also Noblett, "Dru Drury's 'Directions,'" 173–76.

23. Dru Drury to Mr. Hough, March 22, 1762, Drury Letter Book, 13. Drury identified Hough as "Going to Africa with Capt. Johnson as Surgeon" in 1762. Drury's letter also indicates that Hough's vessel was a slave ship bound for Jamaica. There are two voyages commanded by a Captain Johnson and matching Hough's itinerary in the Trans-Atlantic Slave Trade Database: the *Dulce*, commanded by Robert Johnson (voyage ID 90985), and the *Dolphin*, commanded by George Johnson (voyage ID 91005).

24. Dru Drury to Nonus Parke, July 18, 1769, Drury Letter Book, 165.

25. Dru Drury to Nonus Parke, July 18, 1769, Drury Letter Book, 165; and Dru Drury to Mr. Hough, March 22, 1762, Drury Letter Book, 13.

26. Dru Drury to Mr. Hough, March 22, 1762, Drury Letter Book, 12; and Dru Drury to Mr. Richards, November 22, 1769, 177.

27. Drury Letter Book, folio 1v; and "Coleoptera Notebook," Drury Catalogue.

28. Delbourgo, *Collecting the World*, xxviii, 202; Delbourgo, "Listing People"; and Swann, *Curiosities and Texts*, 90.

29. Dru Drury, "Account of Boxes with Instruments &c for catching Insects deliver'd to Different Persons &c," 1766–69, Drury Letter Book, folio 1r–v; "Copy of a List of Boxes delivered on board different Ships since Sept'r 1768," 1768–69, Drury Letter Book, folio 2(a)r; "A List of Boxes deliverd on board different Ships since Sept'r 1768," 1768–69, Drury Letter Book, folio 2(b)r; "Account of Boxes with Instruments &c for catching Insects deliverd to different persons in 1770," 1769–70, Drury Letter Book, folio 3v; and "Account of Boxes with Instruments for collecting Insects deliver'd to different persons in 1771," Drury Letter Book, folio 3r. The ledger at the beginning of Drury's letter book records deliveries of boxes between November 1766 and January 1773.

30. Drury's reliance on a ledger to track his potential collectors was not unique. As historian of science Anke te Heesen observed, "the weighing and recording of debits and credits was a technique commonly used to narrate and order biography and natural history alike." Natural historians occasionally borrowed techniques from mercantile accounting practices. Such borrowing makes particular sense in Drury's case, given his mercantile background. Te Heesen, "Accounting for the Natural World," 238.

31. Drury Letter Book, folio 1r.

32. Dru Drury to Devereux Jarratt, January 18, 1767, Drury Letter Book, 98; and Drury Letter Book, folios 1v–3r.

33. Drury, *Illustrations*, 1:iii.

34. Drury Letter Book, folios 1v–3r.

35. Dru Drury to Capt. Mayle [Male], October 30, 1766, Drury Letter Book, 83.

36. Dru Drury to James Kallender, December 29, 1768, Drury Letter Book, 144.

37. Of the 167 entries, 55 (or 33 percent) in Drury's ledger were for travelers bound to West and West Central Africa. Drury Letter Book, folios 1v–3r. Deane and Cole, *British Economic Growth*, 87. David Eltis notes that even in the British slave trade's busiest year, 1792, the slave trade only accounted for 1.5 percent of British vessels and less than 3 percent of British shipping tonnage. Eltis, *Rise of African Slavery*, 265.

38. Thirty-three of the thirty-six slaving voyages were identified using the Trans-Atlantic Slave Trade Database. Drury Letter Book, folios 1v–3r. Seven of the twelve unknowns were bound for Senegal, so they may have been engaged in Senegambia's gum trade.

39. Drury Letter Book, folios 1v–3r. Twenty-nine of the thirty-three (or 88 percent) of voyages for which the port of origin is known sailed out of London. Four voyages recorded in Drury's ledger began in Liverpool. The port of origin for the remaining twenty voyages is unknown.

40. Rawley, *London, Metropolis*, xi; and Dru Drury to Nonus Parke, November 3, 1770, Drury Letter Book, 219.

41. Dru Drury to Thomas Miller, July 17, 1766, Drury Letter Book, 73.

42. Drury Letter Book, folio 1v; *Slave Voyages: Trans-Atlantic* (voyage ID 77908); Drury, "Directions for Collecting Insects in Foreign Countries," BL; and Noblett, "Dru Drury's 'Directions.'" See also Dru Drury to James Kallender, December 29, 1768, Drury Letter Book, 144; and Dru Drury to Mr. Wollin, March 16, 1767, Drury Letter Book, 103.

43. Dru Drury to William Dalrymple, October 6, 1774, Drury Letter Book, 331.

44. This includes individuals identified in Drury's ledger as well as in the catalog to his collection. Drury Letter Book, folios 1v–3r; and Drury Catalogue.

45. Rediker, *Slave Ship*, 70 (quotation), 59–60, 165; and *Slave Voyages: Trans-Atlantic* (voyage ID 77990).

46. Drury Letter Book, folios 1v–3r.

47. Dru Drury to William Dalrymple, October 6, 1774, Drury Letter Book, 331. See also Rediker, *Slave Ship*, 57–59, 229.

48. Rediker, *Slave Ship*, 9 (quotation), 58–59, 216, 232–39.

49. Drury Letter Book, folio 3v.

50. For the *Good Intent*, see Drury Letter Book, folio 1v. The Trans-Atlantic Slave Trade Database reports the voyage was "Shipwrecked or destroyed, before slaves embarked." *Slave Voyages: Trans-Atlantic* (voyage ID 91395). For Wright, see Drury Letter Book, folio 1v. For Burn, see Drury Letter Book, folio 3v; and *Slave Voyages: Trans-Atlantic* (voyage ID 78031).

51. Drury Letter Book, folio 1r–v; and Drury Catalogue. The two voyages referenced in Drury's ledger appear in Trans-Atlantic Slave Trade Database as voyage ID 77908 and 77980. Wallace likely commanded two subsequent voyages on the *Africa* (voyage ID 24694 and 78050). See *Slave Voyages: Trans-Atlantic*. The estimated number of specimens Wallace collected on his 1767 voyage on the *Africa* is based on Drury's note that he paid £1.1.0 for the specimens and his promise in *Directions* to pay sixpence per specimen.

52. *Slave Voyages: Trans-Atlantic* (voyage ID 77915, 77984); and Drury Letter Book, folio 1r–v.

53. Drury Letter Book, folio 1r–v; and *Slave Voyages: Trans-Atlantic* (voyage ID 77915, 77968, 78276).

54. Drury Letter Book, folio 1r–v; and *Slave Voyages: Trans-Atlantic* (voyage ID 91395, 77950).

55. Dru Drury to Nonus Parke, November 3, 1770, and February 10, 1771, Drury Letter Book, 219 ("fate & success" and "I shall be") and 228 ("with their claws" and "by tearing them"); Drury, "Directions for Collecting Insects in Foreign Countries," 174 ("are quite dead"). See also *Slave Voyages: Trans-Atlantic* (voyage ID 91573).

56. Dru Drury to Nonus Parke, March 9, 1773, Drury Letter Book, 266. See also Drury Catalogue; and *Slave Voyages: Trans-Atlantic* (voyage ID 91574).

57. Drury Letter Book, folios 1v–3r; and Drury Catalogue. Nineteen of Parke's twenty-one specimens came from Calabar, one from Benin, and one is simply labeled "Africa."

58. Drury, *Illustrations of Natural History*, 2:27 ("received this species"), 79 ("from Callabar"). See also "Various Species Notebook," Drury Catalogue, folio 14v; and "Lepidoptera Notebook," Drury Catalogue, folio 8v.

59. Drury Letter Book, folio 3r; and *Slave Voyages: Trans-Atlantic* (voyage ID 91575, 91576, 91594, 91595).

60. Dru Drury to Mrs. Linnecar, September 13, 1772, Drury Letter Book, 252.

61. Dru Drury to Richard Cowley, November 20, 1769, Drury Letter Book, 176.

62. Drury Catalogue. This total does not include the more than 800 specimens gathered by Henry Smeathman, which forms the subject of the next chapter. Unlike the mariners who are the focus of this chapter, Smeathman spent an extended length of time in the region. He lived in Sierra Leone for four years collecting specimens on behalf of a group of patrons that included Drury.

63. Drury Catalogue.

64. Volume 1 of Drury's *Illustrations of Natural History* was published in 1770, but Harris began the engravings for it in 1766, and Drury expected it to be printed in 1768. Noblett, "Publishing by the Author," 69–70.

65. Drury Catalogue. Drury received 85 percent of the West and West Central African specimens in his collection from mariners and other travelers between 1768 and 1773.

66. Westwood, *Illustrations of Exotic Entomology*, 1:iii–v, iii (quotations); and *Superb Collection of Insects*. For an insightful discussion of the dispersal by auction of a larger but contemporary collection, see Tobin, *The Duchess's Shells*, 3–10, 215–45.

67. Drury, *Illustrations of Natural History*, 3:56; Drury Catalogue; Hancock and Douglas, "William Hunter's Goliath Beetle," 224; and Westwood, *Illustrations of Exotic Entomology*, 3:54.

CHAPTER SIX

1. "Extract from Mr. Henry Smeathman's Journal," 15–17, UUL; and Elizabeth Smeathman to John Coakley Lettsom, January 3, 1787, in Pettigrew, *Memoirs of the Life*, 2:256–57, 259.

2. "Extract from Mr. Henry Smeathman's Journal," 15–17. Emphasis in the original.

3. Smeathman, "Proposals for Printing by Subscription"; and Elizabeth Smeathman to John Coakley Lettsom, January 3, 1787, 261.

4. "Extract from Mr. Henry Smeathman's Journal," 17, UUL.

5. Smeathman is most frequently referenced in eighteenth-century scholarship in connection with the British colonization of Sierra Leone. The naturalist's *Plan of a Settlement to be made near Sierra Leona* (1786) is often credited as the inspiration for the British colony that followed. The natural historical expedition that took Smeathman to Sierra Leone fifteen years earlier had received less scholarly attention until recently. Recent works by Alexandra Starr Douglas and Deidre Coleman fill in important gaps in our understanding of Smeathman.

See Coleman, *Henry Smeathman*; Coleman, "Henry Smeathman, the Fly-Catching Abolitionist"; Douglas, "The Making of Scientific Knowledge"; and Douglas, "Natural History, Improvement, and Colonisation."

6. "Extract from Mr. Henry Smeathman's Journal," 16, UUL; and Rodney, *History of the Upper Guinea Coast*, 152.

7. "Extract from Mr. Henry Smeathman's Journal," 17, UUL.

8. Henry Smeathman to John Coakley Lettsom, October 19, 1782, in Lettsom, *Works of John Fothergill*, 3:184–85, 185 ("some lover"). See also Elizabeth Smeathman to Lettsom, January 3, 1787, in Pettigrew, *Memoirs of the Life*, 2:253–59.

9. Elizabeth Smeathman to John Coakley Lettsom, January 3, 1787, in Pettigrew, *Memoirs of Life*, 2:254–55; and Abbot, "Notes on My Life," 28.

10. Henry Smeathman to John Coakley Lettsom, October 19, 1782, in Lettsom, *Works of John Fothergill*, 3:185 (quotation), 184–85.

11. Dru Drury to Henry Smeathman, July 23, 1771, Drury Letter Book; Dru Drury to Henry Smeathman, August 21, 1771, Drury Letter Book, 238; Dru Drury to Henry Smeathman, November 30, 1773, Drury Letter Book, 295–96; Coleman, *Henry Smeathman*, 1; and Tobin, *Duchess's Shells*, 64–65, 73–74.

12. Dru Drury to Henry Smeathman, April 28, 1772, Drury Letter Book, 245–46; and Dru Drury to Henry Smeathman, January 30, 1774, Drury Letter Book, 306–7.

13. Brigham, "Mark Catesby."

14. Tobin, *Duchess's Shells*, 115–43.

15. Dru Drury to Henry Smeathman, May 26, 1774, Drury Letter Book, 318.

16. Dru Drury to Henry Smeathman, May 26, 1774, Drury Letter Book, 318; and Dru Drury to Henry Smeathman, July 4, 1773, Drury Letter Book, 279.

17. "Mr. Smeathman's Letters to Mr. Drury," 34, UUL; and Dru Drury to Henry Smeathman, January 30, 1774, Drury Letter Book, 306.

18. Terrall, "African Indigo," 2–24; Nicolas, "Adanson, the Man," 1–121, esp. 29; and Adanson, *Voyage to Senegal*.

19. Brown, "Empire without America," esp. 87; Dziennik, "'Till these Experiments be Made'"; and Martin, *British West African Settlements*, 57–75. Senegambia remained a British colony for less than twenty years, ending when the French retook Senegal during the American Revolutionary War.

20. Adanson, *Voyage to Senegal*, ix.

21. "Extract from Mr. Smeathman to several Gentlemen," 18, UUL.

22. "Extract from Mr. Smeathman to several Gentlemen," 34, UUL. Cleveland's father was an English slave trader who arrived in the Upper Guinea coast in the mid-eighteenth century. His mother was African from an aristocratic family. Rodney, *A History*, 220; and Coleman, *Henry Smeathman*, 88.

23. In 1773 Smeathman's annual gift included a rather unique addition: a quintessential eighteenth-century scientific object, an "electrical machine." Unfortunately, there is no further reference to Cleveland's electrical machine, leaving us to only wonder what Smeathman's landlord thought of such a gift. Dru Drury to Henry Smeathman, January 28, 1773, Drury Letter Book, 259. Deidre Coleman argues that the gift of the electrical machine marked James Cleveland's elevation to ruler of the Bananas at the death of his brother John. Coleman, *Henry Smeathman*, 101.

24. Brooks, *Landlords and Strangers*, 38–39, 137; Dorjahn and Fyfe, "Landlord and Stranger," 394–96; Mouser, "Landlords-Strangers," 425–30; and Rodney, *A History*, 83–94.

For temporary marriages between African women and European men, see Ipsen, *Daughters of the Trade*.

25. "Extract from Mr. Henry Smeathman's Journal," 24, UUL. As a naturalist primarily interested in entomology, Smeathman felt inadequate to the task of collecting and describing the many new plants he found himself surrounded by in Sierra Leone. He therefore requested that his friends in London help him to recruit a botanical assistant to join him. Anders Berlin, a young Swedish botanist and former student of Carolus Linnaeus, joined Smeathman's household in April 1773. Dru Drury to Henry Smeathman, November 20, 1772, Drury Letter Book, 254.

26. "Extract from Mr. Henry Smeathman's Journal," 24–45, UUL; "Mr. Smeathman's Letters to Mr. Drury," 7, UUL; and Coleman, *Henry Smeathman*, 96–99.

27. "Mr. Smeathman's Letters to Mr. Drury," 10, UUL.

28. Dru Drury to Henry Smeathman, May 26, 1774, Drury Letter Book, 318.

29. "Mr. Smeathman's Letters to Mr. Drury," 3 (quotations; emphasis in the original), UUL; and Ipsen, *Daughters of the Trade*, esp. 1–53. See also Douglas, "Making of Scientific Knowledge."

30. Dru Drury to Henry Smeathman, July 23, 1771, Drury Letter Book, 236; Dru Drury to Henry Smeathman, August 21, 1771, Drury Letter Book, 238; and Dru Drury to Henry Smeathman, March 1, 1772, Drury Letter Book, 242. The recommendation that Smeathman make the Banana Islands his home base may have reflected the belief that certain off-shore islands including the Bananas offered a healthier climate than the West African mainland. Curtin, "'The White Man's Grave,'" 99.

31. Rodney, *A History*, 1–4; Mouser, *Slaving Voyage*, viii–xiii; and Mouser, "Iles de Los as Bulking Center." Slave ship captains were also willing to pay a premium to more quickly purchase captive Africans in order reduce the chances of a slave insurrection and to be the first to reach American markets and therefore benefit from colonists' willingness to pay higher prices.

32. Hancock, *Citizens of the World*, 172–220, 204 ("rendezvous"), 198–99 ("ship trade"); and Rodney, *A History*, 248–52. The island's name is also sometimes spelled as "Bence" or "Bunce."

33. Hancock, *Citizens of the World*, 191–93, 205–6, 213–17.

34. "Mr. Smeathman's Letters to Mr. Drury," 15, UUL; and *Slave Voyages: Trans-Atlantic* (voyage ID 78105). The *Amelia* was owned by Richard Oswald.

35. "Mr. Smeathman's Letters to Mr. Drury," 9–16, 15 (quotations), UUL. See also "Extract from Mr. Smeathman to several Gentlemen," 2–10, 15–17, 19–35, UUL.

36. Smeathman's party included his servant, Hill; his newly arrived botanical assistant, Berlin; the captain of the slave ship who transported Berlin from England; two other slave ship captains; and two agents for British slaving companies. "Mr. Smeathman's Letters to Mr. Drury," 17, UUL; and *Slave Voyages: Trans-Atlantic* (voyage ID 79045 and 78105).

37. "Mr. Smeathman's Letters to Mr. Drury," 18, UUL.

38. "Mr. Smeathman's Letters to Mr. Drury," 23–25, UUL.

39. "Mr. Smeathman's Letters to Mr. Drury," 23–29, 34–35, UUL.

40. "Mr. Smeathman's Letters to Mr. Drury," 26, 29–30, UUL.

41. "Mr. Smeathman's Letters to Mr. Drury," 8–9, 12–13, UUL; and "Extract from Mr. Smeathman to several Gentlemen," 10, UUL.

42. Douglas, "Natural History, Improvement, and Colonisation," 89–92; "Extract from Mr. Smeathman to several Gentlemen," 5–6 ("travelling but by water"), 2 ("great Oxen"), UUL.

43. "Mr. Smeathman's Letters to Mr. Drury," 13, UUL.
44. Dru Drury to Henry Smeathman, November 20, 1772, Drury Letter Book, 254.
45. Dru Drury to Henry Smeathman, July 4, 1773, Drury Letter Book, 278–79.
46. John Fothergill to Carolus Linnaeus, April 4, 1774, in Corner and Booth, *Chain of Friendship*, 409. For more on Fothergill's plan for a second Smeathman-led expedition, see John Fothergill to Henry Smeathman, [contemporary copy by Dru Drury], January 4, 1774, Drury Letter Book, 303; Dru Drury to Henry Smeathman, January 30, 1774, Drury Letter Book, 306; Dru Drury to Henry Smeathman, May 26, 1774, Drury Letter Book, 318; Dru Drury to Henry Smeathman, November 22, 1775, Drury Letter Book, 360; and Dru Drury to John Smeathman, April 11, 1774, Drury Letter Book, 316.
47. Morgan, "Liverpool's Dominance," 25 (table 1.2).
48. Dru Drury to Henry Smeathman, November 30, 1773, Drury Letter Book, 295. See also Dru Drury to John Smeathman, July 19, 1773, Drury Letter Book, 278; Dru Drury to John Smeathman, November 29, 1773, Drury Letter Book, 298; and *Gore's Liverpool Directory*, 52, 68.
49. Dru Drury to Henry Smeathman, November 30, 1773, Drury Letter Book, 295.
50. "Mr. Smeathman's Letters to Mr. Drury," 38, UUL.
51. Terrall, *Catching Nature*, 1–7, 13–43.
52. Smeathman, "Some Account of the Termites," 170.
53. Smeathman, "Some Account of the Termites," 152–54, 153 ("mould" and "white globules").
54. Smeathman, explanation of the plates in "Some Account of the Termites."
55. Smeathman, "Some Account of the Termites." See also Henry Smeathman, January 23, 1781, "Account of the termites," Letters & Papers, vol. 7, no. 188, RS.
56. My focus is on what "Some Account of the Termites" reveals about Smeathman's practices of natural history. For readings of the text as allegories for British colonialism and for its similarities to his plans to colonize Sierra Leone, see Coleman, *Romantic Colonization*, 30–32, 37–49, 58–60; Coleman, "Henry Smeathman, the Fly-Catching Abolitionist"; and Douglas, "Natural History, Improvement, and Colonisation," 124–31.
57. Terrall, *Catching Nature*, 1–20, 2 ("natural history of animals").
58. Smeathman, "Some Account of the Termites," 140.
59. Smeathman, "Some Account of the Termites," 147.
60. Smeathman, "Some Account of the Termites," 176.
61. Smeathman, "Some Account of the Termites," 186.
62. Gómez, *Experiential Caribbean*; Murphy, "Translating the Vernacular"; Parrish, *American Curiosity*, 215–306; Safier, *Measuring the World*, 62–66; and Schiebinger, *Plants and Empire*, 73–104.
63. Smeathman, "Some Account of the Termites," 165.
64. Smeathman, "Some Account of the Termites," 165.
65. "Mr. Smeathman's Letters to Mr. Drury," 40, UUL; and Coleman, "Henry Smeathman and the Natural Economy of Slavery," 138.
66. Andreas Berlin to Carolus Linnaeus, April 15, 1773, L4829, The Linnaean Correspondence—Letters, http://linnaeus.c18.net/Letters; and Henry Smeathman to John Coakley Lettsom, October 1782, in Lettsom, *Works of John Fothergill*, 3:191. Some specimens gathered by Smeathman can be found in modern collections including the Hunterian Museum (Zoology) at the University of Glasgow, the Hunterian Museum at the Royal College of Surgeons of England, and the Linnean Society of London, although these specimens represent

a small fraction of the original collection. The Hunterian Museum at the Royal College of Surgeons includes wet preparations from the comparative anatomist and surgeon John Hunter's dissection of specimens collected by Smeathman. To date, the Hunterian Museum in Glasgow has identified seventy-five insect specimens that Smeathman acquired during his expedition. As the provenance and identity of the remainder of Hunter's insect collection continues to be investigated fully, the number of insects that can be traced to Smeathman's expedition is likely to go up. My many thanks to Jeanne Robinson, curator of entomology at the Hunterian for lending her expertise about the museum's collections. See Smeathman, "Some Account of the Termites," 172; SurgiCat, Royal College of Surgeons of England, http://surgicat.rcseng.ac.uk/; "University Collections: Search the Collections," University of Glasgow, http://collections.gla.ac.uk/#/advancedsearch; and Coleman, *Henry Smeathman*, 22.

67. Dru Drury to Henry Smeathman, November 22, 1775, Drury Letter Book, 360–61, 360 ("My house"). Drury estimated that the collection of plants was so large that it would require one person working for two months to sort them. At least in part, Smeathman's decision to linger in the Caribbean seems motivated by the £20,000 prize Grenadian planters promised to anyone who eradicated the sugarcane ants that plagued the islands.

68. Dru Drury to Henry Smeathman, November 22, 1775, Drury Letter Book, 360–61, 360 ("however hard"); Elizabeth Smeathman to John Coakley Lettsom, January 3, 1787, 260. According to his sister-in-law, Smeathman reserved his most prized specimens to travel with him from the West Indies, but these were captured by an American privateer during his return voyage to Britain in 1779.

69. Smeathman, "Proposals for Printing by Subscription," i–iv; Henry Smeathman, "On Harmattan Winds," November 12, 1785, Manuscripts of the Society for the Promotion of Natural History, LLS; Henry Smeathman, "A Table of the Weather kept at the Bananas on the Coast of Africa in Lat. 8 11 North" (1786), Manuscripts of the Society for the Promotion of Natural History, LLS; Henry Smeathman, [Marine Animalcules and Polyps], Manuscripts of the Society for the Promotion of Natural History, LLS; Henry Smeathman, [Account of the Tarantula of Sierra Leone], 1802, Manuscripts of the Linnean Society, LLS; and Douglas, "Natural History, Improvement, and Colonisation," 117.

70. Harington, *Schizzo on the Genius of Man*, 71–88, 77 ("many other"); and Coleman, "Entertaining Entomology." The literature on scientific and medical lecturing in the eighteenth century is extensive. For a start, Guerrini, "Anatomists and Entrepreneurs"; Schaffer, "Natural Philosophy and Public Spectacle"; and Stewart, *Rise of Public Science*.

71. Smeathman "On Harmattan Winds," 1–2, LLS.

72. Modern historical scholarship largely confirms Smeathman's description of the *Africa*. It was, as he noted, commanded by Capt. John Tittle, who embarked 466 enslaved Africans, only 380 of whom survived the middle passage. *Slave Voyages: Trans-Atlantic* (voyage ID 91495).

73. "Extract from Mr. Henry Smeathman's Journal," 12–14, UUL.

74. It is possible that this was the same vessel. Both were owned by Liverpool merchant Miles Barber and visited his factory at the Îles de Los. However, *Africa* was a popular name for British slave ships.

75. "Mr. Smeathman's Letters to Mr. Drury," 23–29, 26 ("raw"), 27 ("attacked with a fever," "two or three slaves"), UUL.

76. "Mr. Smeathman's Letters to Mr. Drury," 29 ("here they are"), UUL; Coleman, "Henry Smeathman and the Natural Economy of Slavery," 134–47; and Coleman, *Romantic Colonization*, 49–52.

77. "Mr. Smeathman's Letters to Mr. Drury," 14, UUL.

78. "Extract from Mr. Smeathman to several Gentlemen," 18, UUL. See also Coleman, *Romantic Colonization*, 34.

79. Coleman, "Henry Smeathman and the Natural Economy," 138. See also Coleman, *Henry Smeathman*, 176–86, 246–52.

80. Cumberland, "To the Editor," 200.

81. Jonas Hanway, PRO T1/634/2012, Committee Proceedings of April 8, 1786, quoted in Braidwood, *Black Poor and White Philanthropists*, 101; Henry Smeathman to John Hunter, October 15, 1785, Manuscripts of the Society for the Promotion of Natural History, LLS; and Henry Smeathman to John Coakley Lettsom, October 15, 1785, in Pettigrew, *Memoirs of the Life*, 2:281–87.

82. Smeathman, *Plan of a Settlement*.

EPILOGUE

1. Jeanne Robinson, curator of entomology, The Hunterian, personal communication, February 17, 2021; and "University Collections: Search the collections," University of Glasgow, http://collections.gla.ac.uk/#/advancedsearch. One of Smeathman's insect specimens was included in the recent exhibit, "Call and Response: The University of Glasgow and Slavery." The exhibit is part of the university's broader historical slavery initiative; see www.gla.ac.uk/explore/historicalslaveryinitiative/.

2. Hancock, "The Shaping Role of Johan Christian Fabricius," 153, 157–58.

3. Tobin, *The Duchess's Shells*; and *Superb Collection of Insects*. The Smeathman specimens in the Hunterian Museum are unlikely to have been acquired via these auctions, given their timing.

4. See Joseph Ewan's "Appendix: Houstoun Specimens in Lord Petre's Hortus Siccus Stirpium Americanum at Sutro Library, San Francisco," in Ewan, "Plant Collectors in America," 44–47.

5. There is no reason to think that the British were unique in exploiting the trade in human cargo to collect natural historical specimens. Further research is required to determine the scope of collecting by means of the slave trade among the other enslaving powers.

6. Araujo, *Reparations for Slavery*; Beckles, *Britain's Black Debt*; Darity and Mullen, *From Here to Equality*; and Hall, "Doing Reparatory History."

7. Rachel Hatzipanagos, "The 'Decolonization of the American Museum,'" *Washington Post*, October 11, 2018; Hicks, *Brutish Museums*; "Smithsonian Adopts Policy on Ethical Returns," News Release (May 3, 2022), www.si.edu/newsdesk/releases/smithsonian-adopts-policy-ethical-returns; and Parzen, "Knowing Better, Doing Better."

8. The scale of these collections is staggering. Recent estimates suggest that museums in Europe and the United States collectively hold a million Native American skeletons. A 2018 study commissioned by the French government concluded that in French public museums alone, there are at least 90,000 objects of African art. UNESCO estimates between 90 and 95 percent of sub-Saharan African cultural artefacts are housed outside of Africa. Sarr and Savoy, "Restitution of African Cultural Heritage."

9. Laws governing repatriation of human remains vary by nation. In the United Kingdom, for example, the Human Tissue Act 2004 empowers museums to repatriate human remains less than 1,000 years old. In the United States, the Native American Graves Protection Act facilitates the repatriation of Native American remains, but the law does not apply to any

remains for which a connection to living descendants cannot be made. Moreover, it only applies to Native American remains. There is currently no US federal law facilitating the return and reburial of other human remains including those of African Americans whose remains became part of museum collections because of their enslavement. Some institutions, most notably the Smithsonian, have adopted new collecting policies regarding repatriation that go well beyond legal requirements. Delande Justinvil and Chip Colwell, "US Museums Hold the Remains of Thousands of Black People," *The Conversation*, April 15, 2021, https://theconversation.com/us-museums-hold-the-remains-of-thousands-of-black-people-156558; Peggy McGlone, "Human Bones, Stolen Art: Smithsonian Tackles Its 'Problem' Collections," *Washington Post*, July 27, 2002; "Smithsonian Adopts Policy on Ethical Returns"; Redman, *Bone Rooms*; and Jennifer Schuessler, "What Should Museums Do with the Bones of the Enslaved?" *New York Times*, April 20, 2021.

10. Alexandre Antonelli, "The Time Has Come to Decolonize Botanical Gardens Like Kew," *Independent*, June 26, 2020; Das and Lowe, "Nature Read in Black and White," 4–14; Figueiredo and Smith, "Colonial Legacy"; and Nordling, "How Decolonization Could Reshape South African Science." London's Natural History Museum was an early leader in these efforts with the "Slavery and Natural World" project developed for the 2007 bicentenary celebrations of the abolition of the British slave trade. See www.nhm.ac.uk/discover/slavery-and-the-natural-world.html.

BIBLIOGRAPHY

ARCHIVAL AND HERBARIA SOURCES

Biblioteca Nacional de España, Madrid, Spain
 William Toller, "The History of a Voyage to the River of Plate
   & Buenos Ayres from England," (1715), Mss 3039
British Library, London
 Sloane Manuscripts
 South Sea Company Papers
British Museum, London
 Manuscript Catalogues of Sir Hans Sloane's Collections
  Miscellanies Catalogue
Huntington Library, San Marino, California
 Stowe Manuscripts 57
Linnean Society of London Library, London
 Ellis Manuscripts
 Manuscripts of the Society for the Promotion of Natural History
National Archives of the United Kingdom, Richmond
 Board of Trade and Secretaries of State
  America and West Indies, Original Correspondence
   Subseries: Georgia, CO5/670
 Company of Royal Adventurers of England Trading with Africa and Successors
  Records, T70/7, T70/51, T70/53
  Correspondence of the Duke of Chandos, C113/279
  Letters and reports from Cape Coast Castle to the Court of
   the Royal African Company in London, C113/274
Natural History Museum, London
 Dru Drury, "Giving an account of the printing expenses incurred by
   the publication of Illustrations of Natural History (1770–1782)"

Letter-book of Dru Drury, 1761–83
Manuscript Catalogs of Sir Hans Sloane's Collections
    Fossils; Insects; Minerals; and *Vegetable and Vegetable Substances*
    Sloane Herbarium
William Houstoun
    Botanical manuscripts and drawings of plants collected in
        Central America, Jamaica and Cuba, ca. 1730–33
Oxford University Museum of Natural History, Oxford
    "A Catalogue of the Most Capital Assemblage of Insects Probably Ever
        Offered to Public Sale; Consisting of upwards of Eleven Thousand
        different Specimens, collected ... by Mr. Dru Drury ..." [1805]
    "A Catalogue to Exotic Insects in the Collection of Dru Drury, 1784"
Royal Society of London, London
    Classified Papers, 1661–1741
    Journal Books of Scientific Meetings
    Record Book Originals
    Sherard Papers
    Smeathman Images
University of Glasgow Archives and Special Collections, University of Glasgow, Scotland
    Papers of James Douglas, MS Hunter D
University of Oxford, Oxford
    Fielding-Druce Herbarium, Oxford University Herbaria
    Papers of Johann Jacob Dillenius, MSS Sherard 205, Sherardian Library
Uppsala University Library, Uppsala University, Sweden
    "Extract from Mr. Henry Smeathman's Journal, Book 1,"
        [contemporary mss copy], MS D.26, no. 3
    "Extract from Mr. Smeathman to several Gentlemen,"
        [contemporary mss copy], MS D.26, no. 4
    "Mr. Smeathman's Letters to Mr. Drury," [contemporary mss copy], MS D.26, no. 2

DIGITAL SOURCES

Early Modern Letters Online, http://emlo.bodleian.ox.ac.uk/.
The Linnaean Correspondence—Letters, http://linnaeus.c18.net/Letters.
Oxford University Herbaria On-Line Database, https://herbaria.plants.ox.ac.uk/bol/oxford/Explore.
Slave Voyages: The Intra-American Slave Trade Database, www.slavevoyages.org/american/database.
Slave Voyages: The Trans-Atlantic Slave Trade Database, www.slavevoyages.org/voyage/database.

PRINT SOURCES

Abbot, John. "Notes on My Life." *Lepidopterists' News* 2, no. 3 (March 1948): 28–30.
"Account of Books." *Philosophical Transactions* 20, no. 240 (1698): 196–200.
Adanson, Michel. *A Voyage to Senegal, the Isle of Goree, and the River Gambia*. London: J. Nourse in the Strand and W. Johnston in Ludgate-street, 1759.

Aiton, Arthur S. "The Asiento Treaty as Reflected in the Papers of Lord Shelburne." *Hispanic American Historical Review* 8, no. 2 (May 1928): 167–77.

Aiton, William. *Hortus Kewensis; or, A Catalogue of the Plants Cultivated in the Royal Botanic Garden at Kew*. London, 1789.

Allen, Phyllis. "The Royal Society and Latin America as Reflected in the Philosophical Transactions, 1665–1730." *Isis* 37, no. 3/4 (July 1947): 132–38.

"An Account of a Book entituled, *Gazophylacii Naturae & Artis*." *Philosophical Transactions* 27, no. 331 (1710–1712): 342–52.

Araujo, Ana Lucia. *Reparations for Slavery and the Slave Trade: A Transnational and Comparative History*. London: Bloomsbury Academic, 2017.

Aubry, T. *The Sea-Surgeon, or the Guinea Man's Vade Mecum*. London: John Clarke, 1729.

Austin, Daniel F. "Sendera-clandi (*Xenostegia tridentata* (L.) D. F. Austin & Staples, Convolvulaceae): A Medicinal Creeper." *Ethnobotany Research & Applications* 12 (2014): 433–54.

Baker, C. H. Collins, and Muriel I. Baker. *The Life and Circumstances of James Brydges, First Duke of Chandos, Patron of the Liberal Arts*. Oxford: Clarendon Press, 1949.

Baptist, Edward E. *The Half Has Never Been Told: Slavery and the Making of Modern Capitalism*. New York: Basic Books, 2016.

Barcia, Manuel. *The Yellow Demon of Fever: Fighting Disease in the Nineteenth-Century Transatlantic Slave Trade*. New Haven, CT: Yale University Press, 2020.

Barrera-Osorio, Antonio. *Experiencing Nature: The Spanish American Empire and the Early Scientific Revolution*. Austin: University of Texas Press, 2010.

Basalla, George. "The Spread of Western Science." *Science* 156, no. 3775 (1967): 611–22.

Baskes, Jeremy. *Indians, Merchants, and Markets: A Reinterpretation of the Repartimiento and Spanish-Indian Economic Relations in Colonial Oaxaca, 1750–1821*. Stanford, CA: Stanford University Press, 2000.

Baucom, Ian. *Specters of the Atlantic: Finance Capital, Slavery, and the Philosophy of History*. Durham, NC: Duke University Press, 2005.

Beckert, Sven. *Empire of Cotton: A Global History*. New York: Knopf, 2014.

Beckert, Sven, and Seth Rockman. *Slavery's Capitalism: A New History of American Economic Development*. Philadelphia: University of Pennsylvania Press, 2016.

Beckles, Hilary McDonald. *Britain's Black Debt: Reparations for Caribbean Slavery and Native Genocide*. Jamaica: University of the West Indies Press, 2013.

Behrendt, Stephen D. "The Captains in the British Slave Trade, from 1785 to 1807." *Transactions of the Historic Society of Lancashire and Cheshire* 140 (1991): 79–140.

Bleichmar, Daniela. *Visible Empire: Botanical Expeditions and Visual Culture in the Hispanic Enlightenment*. Chicago: University of Chicago Press, 2012.

Borucki, Alex, David Eltis, and David Wheat. "Atlantic History and the Slave Trade to Spanish America." *American Historical Review* 120, no. 2 (April 2015): 433–61.

Bosman, Willem. *A New and Accurate* DESCRIPTION *of the Coast of Guinea*. London, 1705.

Braidwood, Stephen J. *Black Poor and White Philanthropists: London's Blacks and the Foundation of the Sierra Leone Settlement, 1786–1791*. Liverpool: Liverpool University Press, 1994.

Breen, Benjamin. *The Age of Intoxication: Origins of the Global Drug Trade*. Philadelphia: University of Pennsylvania Press, 2019.

———. "The Flip Side of the Pharmacopoeia: Sub-Saharan African Medicines and Poisons in the Atlantic World." In *Drugs on the Page: Pharmacopoeias and Healing Knowledge in the Early Modern Atlantic World*, edited by Matthew James Crawford and Joseph M. Gabriel, 143–59. Pittsburgh: University of Pittsburgh Press, 2019.

Brigham, David R. "Mark Catesby and the Patronage of Natural History in the First Half of the Eighteenth Century." In *Empire's Nature: Mark Catesby's New World Vision*, edited by Amy R. W. Meyers and Margaret Beck Pritchard, 91–146. Chapel Hill: University of North Carolina Press, 1998.

Brock, C. H. "Dru Drury's *Illustrations of Natural History* and the Type Specimen of *Golathus goliatus* Drury." *Journal of the Bibliography of Natural History* 8, no. 3 (1977): 259–65.

Brooks, George E. *Landlords and Strangers: Ecology, Society, and Trade in Western Africa, 1000–1630*. Boulder, CO: Westview Press, 1993.

Brown, Christopher Leslie. "Empire without America: British Plans for Africa in the Era of the American Revolution." In *Abolitionism and Imperialism in Britain, Africa, and the Atlantic*, edited by Derek R. Peterson, 84–100. Athens: Ohio University Press, 2010.

———. *Moral Capital: Foundations of British Abolitionism*. Chapel Hill: University of North Carolina Press, 2006.

———. "The Origins of 'Legitimate Commerce.'" In *Commercial Agriculture, the Slave Trade and Slavery in Atlantic Africa*, edited by Robin Law, Suzanne Schwarz, and Silke Strickrodt, 138–57. Woodbridge, Suffolk: James Currey, 2013.

Brown, Vera Lee. "The South Sea Company and Contraband Trade." *American Historical Review* 31, no. 4 (July 1926): 662–78.

Burkart, Arturo. "Comentario sobre el *Hortus Elthamensis* de Dillenius." *Darwiniana* 11 (November 1957): 367–406.

Candler, Allen D., ed. *The Minutes of the Common Council of the Trustees for Establishing the Colony of Georgia in America*. Vol. 2 of *The Colonial Records of the State of Georgia*. Atlanta: Franklin Printing and Publishing, 1904.

———, ed. *Original Papers, Correspondence, Trustees, General Oglethorpe and Others, 1735–1737*. Vol. 21 of *The Colonial Records of the State of Georgia*. Atlanta: Franklin Printing and Publishing, 1910.

Candler, Allen D., William J. Northen, and Lucian Lamar Knight, eds. *Original Papers, Correspondence, Trustees, General Oglethorpe and Others, 1737–1739*. Vol. 22, pt. 1 of *The Colonial Records of the State of Georgia*. Atlanta: Franklin Printing and Publishing, 1913.

Cagle, Hugh. *Assembling the Tropics: Science and Medicine in Portugal's Empire, 1450–1700*. Cambridge: Cambridge University Press, 2018.

Cañizares-Esquerra, Jorge. Introduction to *Science in the Spanish and Portuguese Empires, 1500–1800*, edited by Daniela Bleichmar, Paula De Vos, Kristin Huffine, and Kevin Sheehan, 1–5. Stanford, CA: Stanford University Press, 2009.

Carney, Judith A., and Richard Nicholas Rosomoff. *In the Shadow of Slavery: Africa's Botanical Legacy in the Atlantic World*. Berkeley: University of California Press, 2009.

Carroll, Patrick James. *Blacks in Colonial Veracruz: Race, Ethnicity, and Regional Development*. 2nd ed. Austin: University of Texas Press, 2001.

Catesby, Mark. *The Natural History of Carolina, Florida and the Bahama Islands*. London, 1731.

Chakrabarti, Pratik. *Materials and Medicine: Trade, Conquest and Therapeutics in the Eighteenth Century*. Manchester: Manchester University Press, 2010.

Chamberlain, E. N. "The Influence of the Slave Trade on Liverpool Medicine." In *Atti del XIVo congresso internazionale di storia della medicina sotto l'alto patronato del Presidente della Republica, Roma-Salerno, settembre 1954*, 2:768–73. Rome: International Congress of the History of Medicine, 1956.

Christopher, Emma. *Slave Ship Sailors and Their Captive Cargoes, 1730–1807*. Cambridge: Cambridge University Press, 2006.

Church, Leslie F. *Oglethorpe: A Study of Philanthropy in England and Georgia.* London: Epworth, 1932.

Clokie, Hermia Newman. *An Account of the Herbaria of the Department of Botany in the University of Oxford.* Oxford: Oxford University Press, 1964.

Coleman, Deirdre. "Entertaining Entomology: Insects and Insect Performers in the Eighteenth Century." *Eighteenth-Century Life* 30, no. 3 (Fall 2006): 107–34.

———. "Henry Smeathman and the Natural Economy of Slavery." In *Slavery and the Cultures of Abolition*, edited by Brycchan Carey and Peter J. Kitson, 130–49. Cambridge: D. S. Brewer, 2007.

———. *Henry Smeathman, the Flycatcher.* Liverpool: Liverpool University Press, 2018.

———. "Henry Smeathman, the Fly-Catching Abolitionist." In *Discourses of Slavery and Abolition: Britain and Its Colonies, 1760–1838*, edited by Brycchan Carey, Markman Ellis, and Sara Salih, 144–54. Basingstoke, Hampshire: Palgrave Macmillan, 2004.

———. *Romantic Colonization and British Anti-Slavery.* Cambridge: Cambridge University Press, 2004.

Coleman, Kenneth. *Colonial Georgia: A History.* New York: Charles Scribner's Sons, 1976.

Coleman, Kenneth, and Milton Ready, eds. *Original Papers, Correspondence to the Trustees, James Oglethorpe, and Others, 1732–1735.* Vol. 20 of *The Colonial Records of the State of Georgia.* Athens: University of Georgia Press, 1982.

———. *Trustees' Letter Book, 1732–1738.* Vol. 29 of *The Colonial Records of the State of Georgia.* Athens: University of Georgia Press, 1985.

Cook, Harold J. "The Cutting Edge of a Revolution? Medicine and Natural History Near the Shores of the North Sea." In *Renaissance and Revolution: Humanists, Scholars, Craftsmen and Natural Philosophers in Early Modern Europe*, edited by J. V. Field and Frank A. J. L. James, 45–61. Cambridge: Cambridge University Press, 1993.

———. "Markets and Cultures: Medical Specifics and the Reconfiguration of the Body in Early Modern Europe." *Transactions of the Royal Historical Society*, 6th ser., 21 (2011): 123–45.

———. *Matters of Exchange: Commerce, Medicine, and Science in the Dutch Golden Age.* New Haven, CT: Yale University Press, 2007.

———. "Physicians and Natural History." In *Cultures of Natural History*, edited by N. Jardine, J. A. Secord, and E. C. Spary, 91–105. Cambridge: Cambridge University Press, 1996.

Corner, Betsy C., and Christopher C. Booth, eds. *Chain of Friendship: Selected Letters of Dr. John Fothergill of London, 1735–1780.* Cambridge, MA: Belknap Press of Harvard University Press, 1971.

Crawford, Matthew James, and Joseph M. Gabriel, eds. "Introduction: Thinking with Pharmacopoeias." In *Drugs on the Page: Pharmacopoeias and Healing Knowledge in the Early Modern Atlantic World*, edited by Matthew James Crawford and Joseph M. Gabriel, 3–16. Pittsburgh: University of Pittsburgh Press, 2019.

Cromwell, Jesse. *The Smugglers' World: Illicit Trade and Atlantic Communities in Eighteenth-Century Venezuela.* Chapel Hill: University of North Carolina Press, 2018.

Cumberland, George. "To the Editor of the Monthly Magazine, Mr. Cumberland's Plan for the Protection and Restoration of Females." *Monthly Magazine* 37 (April 1, 1814): 199–203.

Curran, Andrew S. *The Anatomy of Blackness: Science and Slavery in an Age of Enlightenment.* Baltimore: Johns Hopkins University Press, 2011.

Curtin, P. D. "'The White Man's Grave': Image and Reality, 1780–1850." *Journal of British Studies* 1, no. 1 (November 1961): 94–110.

Daaku, Kwame Yeboa. *Trade and Politics on the Gold Coast, 1600–1720*. Oxford: Clarendon, 1970.

Dandy, J. E. *The Sloane Herbarium: An Annotated List of the* Horti Sicci *Composing It; with Biographical Accounts of the Principal Contributors*. London: Trustees of the British Museum, 1958.

Darity, William A., Jr., and A. Kirsten Mullen. *From Here to Equality: Reparations for Black Americans in the Twenty-First Century*. Chapel Hill: University of North Carolina Press, 2020.

Das, Subhadra, and Miranda Lowe. "Nature Read in Black and White: Decolonial Approaches to Interpreting Natural History Collections." *Journal of Natural Science Collections* 6 (2018): 4–14.

Daston, Lorraine. "Type Specimens and Scientific Memory." *Critical Inquiry* 31, no. 1 (Autumn 2004): 153–82.

Daston, Lorraine, and Peter Galison. *Objectivity*. New York: Zone, 2007.

Davies, K. G. *The Royal African Company*. New York: Atheneum, 1970.

Davis, David Brion. *Inhuman Bondage: The Rise and Fall of Slavery in the New World*. Oxford: Oxford University Press, 2006.

Deane, Phyllis, and W. A. Cole. *British Economic Growth, 1688–1959: Trends and Structure*. Cambridge: Cambridge University Press, 1962.

Delbourgo, James. *Collecting the World: Hans Sloane and the Origins of the British Museum*. Cambridge, MA: Harvard University Press, 2017.

———. "Listing People." *Isis* 103, no. 4 (2012): 735–42.

Delbourgo, James, and Nicholas Dew, eds. *Science and Empire in the Atlantic World*. New York: Routledge, 2008.

Dew, Nicholas. "Scientific Travel in the Atlantic World: The French Expedition to Gorée and the Antilles, 1681–1683." *British Journal for the History of Science* 43, no. 1 (March 2010): 1–17.

Dewhurst, Kenneth. *Thomas Dover's Life and Legacy*. Metuchen, NJ: Scarecrow, 1974.

Dewhurst, Kenneth, and Rex Doublet. "Thomas Dover and the South Sea Company." *Medical History* 18, no. 2 (1974): 107–21.

Donnan, Elizabeth. "The Early Days of the South Sea Company, 1711–1718." *Journal of Economic and Business History* 2 (May 1930): 419–50.

Dorjahn, V. R., and Christopher Fyfe. "Landlord and Stranger: Change in Tenancy Relations in Sierra Leone." *Journal of African History* 3, no. 3 (1962): 391–97.

Dorner, Zachary. *Merchants of Medicines: The Commerce and Coercion of Health in Britain's Long Eighteenth Century*. Chicago: University of Chicago Press, 2020.

Douglas, Alexandra Starr. "The Making of Scientific Knowledge in an Age of Slavery: Henry Smeathman, Sierra Leone and Natural History." *Journal of Colonialism and Colonial History* 9, no. 3 (Winter 2008).

———. "Natural History, Improvement, and Colonisation: Henry Smeathman and Sierra Leone in the Late Eighteenth Century." PhD diss., Royal Holloway, University of London, 2004.

Draper, Nicholas. *The Price of Emancipation: Slave Ownership, Compensation and British Society at the End of Slavery*. Cambridge: Cambridge University Press, 2010.

Drayton, Richard. *Nature's Government: Science, Imperial Britain, and the "Improvement" of the World*. New Haven, CT: Yale University Press, 2000.

Drury, Dru. *Illustrations of Natural History*. 3 vols. London, 1770–82.

Dziennik, Matthew P. "'Till these Experiments be Made': Senegambia and British Imperial Policy in the Eighteenth Century." *English Historical Review* 130, no. 546 (2015): 1132–61.

Eltis, David. *The Rise of African Slavery in the Americas*. Cambridge: Cambridge University Press, 2000.

Ewan, Joseph. "Plant Collectors in America: Backgrounds for Linnaeus." In *Essays in Biohistory, and Other Contributions Presented by Friends and Colleagues to Frans Verdoorn on the Occasion of His 60th Birthday*, edited by Pieter Smit and R. J. Ch. V. ter Laage, 19–54. Utrecht: International Association for Plant Taxonomy, 1970.

Fabian, Ann. *The Skull Collectors: Race, Science, and America's Unburied Dead*. Chicago: University of Chicago Press, 2010.

Fett, Sharla. *Working Cures: Healing, Health, and Power on Southern Slave Plantations*. Chapel Hill: University of North Carolina Press, 2002.

Figueiredo, Estrela, and Gideon F. Smith. "The Colonial Legacy in African Plant Taxonomy: Biological Types and Why We Need Them." *South African Journal of Science* 106, no. 3 (2010): 1–4.

Finucane, Adrian. *The Temptations of Trade: Britain, Spain, and the Struggle for Empire*. Philadelphia: University of Pennsylvania Press, 2016.

Fyfe, Christopher. *A History of Sierra Leone*. Oxford: Oxford University Press, 1962.

Gänger, Stefanie. "World Trade in Medicinal Plants from Spanish America, 1717–1815." *Medical History* 59, no. 1 (2015): 44–62.

Golinski, Jan. *Making Natural Knowledge: Constructivism and the History of Science*. Cambridge: Cambridge University Press, 1998.

Gómez, Pablo F. *The Experiential Caribbean: Creating Knowledge and Healing in the Early Modern Atlantic*. Chapel Hill: University of North Carolina Press, 2017.

Gore's Liverpool Directory, For the Year 1774; An Alphabetical List of the Merchants, Tradesmen, and Principal Inhabitants of the Town of Liverpool. Liverpool, 1774.

Gould, Eliga H. "Entangled Histories, Entangled Worlds: The English-Speaking Atlantic as a Spanish Periphery." *American Historical Review* 112, no. 3 (June 2007): 764–86.

Govier, Mark. "The Royal Society, Slavery and the Island of Jamaica: 1660–1700." *Notes and Records of the Royal Society of London* 53, no. 2 (May 1999): 203–17.

Grahn, Lance R. "Cartagena and Its Hinterland in the Eighteenth Century." In *Atlantic Port Cities: Economy, Culture, and Society in the Atlantic World, 1650–1850*, edited by Franklin W. Knight and Peggy K. Liss, 168–95. Knoxville: University of Tennessee Press, 1991.

Greene, Jack P. "Travails of an Infant Colony: The Search for Viability, Coherence, and Identity in Colonial Georgia." In *Forty Years of Diversity: Essays on Colonial Georgia*, edited by Harvey H. Jackson and Phinizy Spalding, 278–310. Athens: University of Georgia Press, 1984.

Greenfield, Amy Butler. *A Perfect Red: Empire, Espionage, and the Quest for the Color of Desire*. New York: Harper, 2005.

Grew, Nehemjah. *Musaeum Regalis Societatis*. London, 1681.

Grieve, Hilda. *A Transatlantic Gardening Friendship, 1694–1777*. Chelmsford, MA: Historical Association, Essex Branch, 1981.

Guerrini, Anita. "Anatomists and Entrepreneurs in Early Eighteenth-Century London." *Journal of the History of Medicine and Allied Sciences* 59, no. 2 (April 2004): 219–39.

Hair, P. E. H., and Robin Law. "The English in Western Africa to 1700." In *The Origins of Empire: British Overseas Enterprise to the Close of the Seventeenth Century*, edited by Nicholas Canny, 241–63. Vol. 1 of *The Oxford History of the British Empire*. Oxford: Oxford University Press, 1998.

Hall, Catherine. "Doing Reparatory History: Bringing 'Race' and Slavery Home." *Race & Class* 60, no. 1 (2018): 3–21.

Hall, Catherine, Nicholas Draper, Keith McClelland, Katie Donington, and Rachel Lang. *Legacies of British Slave-ownership: Colonial Slavery and the Formation of Victorian Britain.* Cambridge: Cambridge University Press, 2014.

Hall, Gwendolyn Midlo. *Slavery and African Ethnicities in the Americas: Restoring the Links.* Chapel Hill: University of North Carolina Press, 2005.

Halley, Edmond. "The Longitude of Carthagena in America." *Philosophical Transactions* 32, no. 375 (1722–23): 237–38.

Hancock, David. *Citizens of the World: London Merchants and the Integration of the British Atlantic Community, 1735–1785.* Cambridge: Cambridge University Press, 1997.

———. *Oceans of Wine: Madeira and the Emergence of American Trade and Taste.* New Haven, CT: Yale University Press, 2009.

Hancock, E. Geoffrey. "The Shaping Role of Johan Christian Fabricius (1745–1808): William Hunter's Insect Collection and Entomology in Eighteenth-Century London." In *William Hunter's World: The Art and Science of Eighteenth-Century Collecting*, edited by E. G. Hancock, N. Pearce, and M. Campbell, 151–64. Abingdon, Oxfordshire: Ashgate, 2015.

Hancock, E. Geoffrey, and A. Starr Douglas. "William Hunter's Goliath Beetle, *Goliathus goliatus* (Linnaeus, 1771), Revisited." *Archives of Natural History* 36, no. 2 (2009): 218–30.

Harington, Edward. *A Schizzo on the Genius of Man: In Which, Among Various Subjects, the Merit of Mr. Thomas Barker, the Celebrated Young Painter of Bath, Is Particularly Considered, and His Pictures Reviewed. By the Author of An Excursion from Paris to Fontainbleau for the Benefit of the Bath Casualty Hospital.* Bath: R. Cruttwell, 1793.

Herschthal, Eric. *The Science of Abolition: How Slaveholders Became the Enemies of Progress.* New Haven, CT: Yale University Press, 2021.

Hevly, Richard H. "Nomenclatural History and Typification of Martynia and Proboscidea (Martyniaceae)." *Taxon* 18, no. 5 (October 1969): 527–34.

Hicks, Dan. *The Brutish Museums: The Benin Bronzes, Colonial Violence and Cultural Restitution.* London: Pluto, 2020.

Hogarth, Rana A. *Medicalizing Blackness: Making Racial Difference in the Atlantic World, 1780–1840.* Chapel Hill: University of North Carolina Press, 2017.

Holland, James W. "The Beginning of Public Agricultural Experimentation in America: The Trustees' Garden in Georgia." *Agricultural History* 12, no. 3 (July 1938): 271–98.

Houstoun, James. *The Works of James Houstoun, M.D.* London, 1753.

Houstoun, William. "An Account of the Contrayerva, by Mr. William Houstoun, Surgeon in the Service of the Honourable South-Sea Company." *Philosophical Transactions* 37, no. 421 (1731): 195–98.

———. *Reliquae Houstounianae: Seu Plantarum in America Meridionali a Gulielmo Houstoun M.D. R.S.S. Collectarum Icones Manu Propria Aere Incisae; cum Descriptionibus e Schedis ejusdem in Bibliotheca Josephi Banks, Baroneti, R.S.P. Asservatis.* London, 1781.

Hunt, Arnold. "Under Sloane's Shadow: The Archive of James Petiver." In *Archival Afterlives: Life, Death, and Knowledge-Making in Early Modern British Scientific and Medical Archives*, edited by Vera Keller, Anna Marie Roos, and Elizabeth Yale, 194–221. Leiden: Brill, 2018.

Huxtable, Sally-Anne, Corinne Fowler, Christo Kefalas, and Emma Slocombe. "Interim Report on the Connections between Colonialism and Properties Now in the Care of the National Trust, Including Links with Historic Slavery." National Trust, September 2020.

Iannini, Christopher P. *Fatal Revolutions: Natural History, West Indian Slavery, and the Routes of American Literature*. Chapel Hill: University of North Carolina Press, 2012.

"Inquires for Suratte and Other Parts of the East-Indies; Inquiries for Persia; Inquiries for Virginia and the Bermudas; for Guaiana and Brasil." *Philosophical Transactions* 2, no. 23 (1666–67): 415–22.

Ipsen, Pernille. *Daughters of the Trade: Atlantic Slavers and Interracial Marriage on the Gold Coast*. Philadelphia: University of Pennsylvania Press, 2015.

Isaac, Grant E., and William A. Kerr. "Bioprospecting or Biopiracy? Intellectual Property and Traditional Knowledge in Biotechnology Innovation." *Journal of World Intellectual Property* 7, no. 1 (January 2004): 35–52.

Jarvis, Charles E. "'The Most Common Grass, Rush, Moss, Fern, Thistles, Thorns or Vilest Weed You Can Find': James Petiver's Plants." *Notes and Records: The Royal Society Journal of the History of Science* 74, no. 2 (June 2020): 303–28.

Jenkins, Susan. *Portrait of a Patron: The Patronage and Collecting of James Brydges, 1st Duke of Chandos (1674–1744)*. Hampshire: Ashgate, 2007.

Johnson, Walter. *River of Dark Dreams: Slavery and Empire in the Cotton Kingdom*. Cambridge, MA: Harvard University Press, 2013.

Johnston, Edith Duncan. "Doctor William Houstoun: Botanist." *Georgia Historical Quarterly* 25, no. 4 (December 1941): 325–39.

Kellman, Jordan. "Nature, Networks, and Expert Testimony in the Colonial Atlantic: The Case of Cochineal." *Atlantic Studies* 7, no. 4 (December 2010): 383–86.

Kenny, Stephen C. "The Development of Medical Museums in the Antebellum American South: Slave Bodies in Networks of Anatomical Exchange." *Bulletin of the History of Medicine* 87, no. 1 (2013): 32–62.

Klooster, Wim. "Inter-Imperial Smuggling in the Americas, 1600–1800." In *Soundings in Atlantic History: Latent Structures and Intellectual Currents, 1500–1830*, edited by Bernard Bailyn and Patricia L. Denault, 141–80. Cambridge, MA: Harvard University Press, 2009.

Koot, Christian. *Empire at the Periphery: British Colonists, Anglo-Dutch Trade, and the Development of the British Atlantic, 1621–1713*. New York: New York University Press, 2011.

Kriger, Colleen E. "'Our indico designe': Planting and Processing Indigo for Export, Upper Guinea Coast, 1684–1702." In *Commercial Agriculture, the Slave Trade and Slavery in Atlantic Africa*, edited by Robin Law, Suzanne Schwarz, and Silke Strickrodt, 98–115. Woodbridge, Suffolk: James Currey, 2013.

Kriz, Kay Dian. "Curiosities, Commodities, and Transplanted Bodies in Hans Sloane's 'Natural History of Jamaica.'" *William and Mary Quarterly*, 3rd series, 57, no. 1 (January 2000): 35–78.

Lamb, D. P. "Volume and Tonnage of the Liverpool Slave Trade, 1772–1807." In *Liverpool, the African Slave Trade, and Abolition*, edited by Roger Anstey and P.E.H. Hair, 91–112. Bristol: Historic Society of Lancashire and Cheshire, 1976.

Lambert, David. *Mastering the Niger: James MacQueen's African Geography and the Struggle over Atlantic Slavery*. Chicago: University of Chicago Press, 2013.

Law, Robin, ed. *The English in West Africa, 1691–1699*. Part 3 of *The Local Correspondence of the Royal African Company of England, 1681–1699*. Oxford: Oxford University Press, 2007.

———. "King Agaja of Dahomey, the Slave Trade, and the Question of West African Plantations: The Embassy of Bulfinch Lambe and Adomo Tomo to England, 1726–32." *Journal of Imperial and Commonwealth History* 19, no. 2 (1991): 137–63.

———. "'There's Nothing Grows in the West Indies But Will Grow Here': Dutch and English Projects of Plantation Agriculture on the Gold Coast, 1650s–1780s." In *Commercial Agriculture, the Slave Trade and Slavery in Atlantic Africa*, edited by Robin Law, Suzanne Schwarz, and Silke Strickrodt, 116–37. Woodbridge, Suffolk: James Currey, 2013.
Lettsom, John Coakley. *The Works of John Fothergill, M.D.* 3 vols. London: Charles Dilly, 1784.
Linnaeus, Carolus. *Systema Naturae*. 10th ed. 2 vols. Stockholm, 1758.
Livingstone, David N. *Putting Science in Its Place: Geographies of Scientific Knowledge*. Chicago: University of Chicago Press, 2003.
Lucas, A. M., and P. J. Lucas. "Natural History 'Collectors': Exploring the Ambiguities." *Archives of Natural History* 41, no. 1 (2014): 63–74.
MacGregor, Neil. *A History of the World in 100 Objects*. London: Penguin, 2010.
MacLeod, Roy, ed. *Nature and Empire: Science and the Colonial Enterprise*. Osiris, 2nd series, 15 (2000).
Margócsy, Dániel. *Commercial Visions: Science, Trade, and Visual Culture in the Dutch Golden Age*. Chicago: University of Chicago Press, 2014.
Martin, Eveline C. *The British West African Settlements, 1750–1821: A Study in Local Administration*. London: Longmans, 1927.
Martin, Franklin W., and Ruth M. Ruberté. *Edible Leaves of the Tropics*. 2nd ed. Mayagüez, Puerto Rico: Antillian College Press, 1979.
Martyn, John. *Historia plantarum rariorum*. London, 1728–37.
Maydom, Katrina Elizabeth. "James Petiver's Apothecary Practice and the Consumption of American Drugs in Early Modern London." *Notes and Records: The Royal Society Journal of the History of Science* 74, no. 2 (June 2020): 213–38.
McClellan III, James. *Colonialism and Science: Saint Domingue in the Old Regime*. Chicago: University of Chicago Press, 2010.
McPherson, Robert, ed. *Manuscripts of The Earl of Egmont: Diary of The First Earl of Egmont (Viscount Percival)*. Vol. 3, 1739–1747. London: H. M. Stationary Office, 1923.
Melville, Lewis. *The South Sea Bubble*. 1921. Reprint, New York: B. Franklin, 1968.
Merson, John. "Bio-Prospecting or Bio-Piracy: Intellectual Property Rights and Biodiversity in a Colonial and Postcolonial Context." *Nature and Empire: Science and the Colonial Enterprise*. Vol. 15 of *Osiris*, 2nd series (2000): 282–96.
Miller, Joseph. *Botanicum officinale; or a compendium herbal; giving an account of all such plants as are now used in the practice of physick. With their descriptions and virtues*. London, 1722.
Miller, Philip. *Figures of the Most Beautiful, Useful, and Uncommon Plants Described in the Gardeners Dictionary*. 2 vols. London: 1760.
———. *The Gardeners Dictionary; Containing the Methods of Cultivating and Improving All Sorts of Trees, Plants, and Flowers*. 8th ed. London, 1768.
Mitchell, Matthew David. "'Legitimate Commerce' in the Eighteenth Century: The Royal African Company of England under the Duke of Chandos, 1720–1726." *Enterprise & Society* 14, no. 3 (September 2013): 544–78.
Morgan, Kenneth. "Liverpool's Dominance in the British Slave Trade, 1740–1807." In *Liverpool and Transatlantic Slavery*, edited by David Richardson, Suzanne Schwarz, and Anthony Tibbles, 14–42. Liverpool: Liverpool University Press, 2007.
Mouser, Bruce L. "Iles de Los as Bulking Center in the Slave Trade, 1750–1800." *Outre-Mers: Revue d'histoire* 83, no. 313 (1996): 77–91.
———. "Landlords-Strangers: A Process of Accommodation and Assimilation." *International Journal of African Historical Studies* 8, no. 3 (1975): 425–40.

———. *A Slaving Voyage to Africa and Jamaica: The Log of the* Sandown, *1793–1794*. Bloomington: Indiana University Press, 2002.

Murphy, Kathleen S. "Collecting Slave Traders: James Petiver, Natural History, and the British Slave Trade." *William and Mary Quarterly*, 3rd series, 70, no. 4 (October 2013): 637–70.

———. "James Petiver's 'Kind Friends' and 'Curious Persons' in the Atlantic World: Commerce, Colonialism, and Collecting." *Notes and Records: The Royal Society Journal of the History of Science* 74, no. 2 (June 2020): 259–74.

———. "A Slaving Surgeon's Collection: The Pursuit of Natural History through the British Slave Trade to Spanish America." In *Curious Encounters: Voyaging, Collecting, and Making Knowledge in the Long Eighteenth Century*, edited by Adriana Craciun and Mary Terrall, 138–58. Toronto: University of Toronto Press, 2019.

———. "Translating the Vernacular: Indigenous and African Knowledge in the Eighteenth-Century British Atlantic." *Atlantic Studies* 8, no. 1 (March 2011): 29–48.

Mustakeem, Sowande' M. *Slavery at Sea: Terror, Sex, and Sickness in the Middle Passage*. Urbana: University of Illinois Press, 2016.

Nelson, George H. "Contraband Trade under the Asiento, 1730–1739." *American Historical Review* 51, no. 1 (October 1945): 55–67.

Newman, Simon P. *A New World of Labor: The Development of Plantation Slavery in the British Atlantic*. Philadelphia: University of Pennsylvania Press, 2013.

Newton, Joshua D. "Naval Power and the Province of Senegambia, 1758–1779." *Journal for Maritime Research* 15, no. 2 (2013): 129–47.

———. "Slavery, Sea Power and the State: The Royal Navy and the British West African Settlements, 1748–1756." *Journal of Imperial and Commonwealth History* 41, no. 2 (2013): 171–93.

Nicolas, Jean-Paul. "Adanson, the Man." In *Adanson: The Bicentennial of Michel Adanson's Familles des plantes, Part I*, edited by George H. M. Lawrence, 1–121. 2 parts. Pittsburgh: Hunt Botanical Library, 1963.

Noblett, William. "Dru Drury's 'Directions for Collecting Insects in Foreign Countries.'" *Bulletin of the Amateur Entomologists' Society* 44 (November 1985): 170–78.

———. "Publishing by the Author: A Case Study of Dru Drury's 'Illustrations of Natural History' (1770–82)." *Publishing History* 23 (1988): 67–94.

Nordling, Linda. "How Decolonization Could Reshape South African Science." *Nature* (February 7, 2018): 159–62.

Norton, Marcy. *Sacred Gifts, Profane Pleasures: A History of Tobacco and Chocolate in the Atlantic World*. Ithaca, NY: Cornell University Press, 2008.

Ogborn, Miles. "Talking Plants: Botany and Speech in Eighteenth-Century Jamaica." *History of Science* 51, no. 3 (September 2013): 251–81.

Oglethorpe, James Edward. "Some Account of the Designs of the Trustees for Establishing the Colony of Georgia in America." 1733. Reprinted in *A Brief Account of the Establishment of the Colony of Georgia under Gen. James Oglethorpe*. Washington, DC, 1835, available at http://memory.loc.gov.

O'Malley, Gregory E. *Final Passages: The Intercolonial Slave Trade of British America, 1619–1807*. Chapel Hill: University of North Carolina Press, 2014.

Osseo-Asare, Abena Dove. *Bitter Roots: The Search for Healing Plants in Africa*. Chicago: University of Chicago Press, 2014.

Palmer, Colin A. *Human Cargoes: The British Slave Trade to Spanish America, 1700–1739*. Urbana: University of Illinois Press, 1981.

Parrish, Susan Scott. *American Curiosity: Cultures of Natural History in the Colonial British Atlantic World*. Chapel Hill: University of North Carolina Press, 2006.

Parsons, Christopher M., and Kathleen S. Murphy. "Ecosystems under Sail: Specimen Transport in the Eighteenth-Century French and British Atlantics." *Early American Studies* 10 (Fall 2012): 503–39.

Parzen, Micah D. "Knowing Better, Doing Better: The San Diego Museum of Man Takes a Holistic Approach to Decolonization." *Museum Magazine* (January/February 2020).

Petiver, James. "An Account of a Book: Musei Petiveriani Centuria Prima." *Philosophical Transactions* 19, no. 224 (January 1697): 393–400.

———. *Brief Directions for the Easie Making, and Preserving Collections of all Natural Curiosities*. London, [1709?].

———. "A Catalogue of some *Guinea-Plants*, with their *Native Names* and *Virtues*; Sent to *James Petiver*, Apothecary, and Fellow of the Royal Society; with his *Remarks* on Them. Communicated in a Letter to Dr. *Hans Sloane*. Secret. Reg. Soc." *Philosophical Transactions* 19, no. 232 (September 1697): 677–86.

———. *Gazophylacii Naturae & Artis, Decas Prima & Secunda*. London, 1702–4.

———. *Gazophylacii Naturae & Artis, Decas Septima and Octava*. London, 1706–11.

———. *Gazophylacii Naturae & Artis, Decas Tertia*. London, 1704.

———. *Hortus Peruvianus Medicinalis: Or, The South-Sea Herbal*. London, 1715.

———. *Musei Petiveriani, Centuria Nona & Decima*. London, 1703.

———. *Musei Petiveriani, Centuria Prima*. London, 1695.

———. *Musei Petiveriani, Centuria Quarta & Quinta*. London, 1699.

———. *Musei Petiveriani, Centuria Secunda & Tertia*. London, 1698.

Pettigrew, Thomas Joseph. *Memoirs of the Life and Writings of the late John Coakley Lettsom . . . With a Selection from his Correspondence*. 3 vols. London, 1817.

Pettigrew, William. *Freedom's Debt: The Royal African Company and the Politics of the Atlantic Slave Trade*. Chapel Hill: University of North Carolina Press, 2013.

Pickering, Victoria Rose Margaret. "Putting Nature in a Box: Hans Sloane's 'Vegetable Substances' Collection." PhD diss., Queen Mary, University of London, 2016.

Pinney, Thomas. *A History of Wine in America*. Vol. 1, *From the Beginnings to Prohibition*. Berkeley: University of California Press, 2007.

Plukenet, Leonard. *Almagestum Botanicum*. London, 1696.

Porter, R. "The Crispe Family and the African Trade." *Journal of African History* 9, no. 1 (1968): 57–77.

Prado, Fabrício. *Edge of Empire: Atlantic Networks and Revolution in Bourbon Rio de la Plata*. Berkeley: University of California Press, 2015.

Pratt, Mary Louise. *Imperial Eyes: Travel Writing and Transculturation*. London: Routledge, 1992.

Raj, Kapil. *Relocating Modern Science: Circulation and the Construction of Knowledge in South Asia and Europe, 1650–1900*. New York: Palgrave Macmillan, 2007.

Rand, Isaac. *Horti Medici Chelseiani index compendiarius*. London, 1739.

Rauschenberg, Roy A. "John Ellis, Royal Agent for West Florida." *Florida Historical Quarterly* 62, no. 1 (1983): 1–24.

Rawley, James A. *London, Metropolis of the Slave Trade*. Columbia: University of Missouri Press, 2003.

Rea, Robert R. *Major Robert Farmar of Mobile*. Tuscaloosa: University of Alabama Press, 1990.

Rediker, Marcus. *The Slave Ship: A Human History*. New York: Penguin, 2007.
Redman, Samuel L. *Bone Rooms: From Scientific Racism to Human Prehistory in Museums*. Cambridge, MA: Harvard University Press, 2016.
Richardson, David. "The British Empire and the Atlantic Slave Trade." In *The Eighteenth Century*, edited by P. J. Marshall, 440–64. Vol. 2 of *The Oxford History of the British Empire*. Oxford: Oxford University Press, 1998.
Rigby, Nigel. "The Politics and Pragmatics of Seaborne Plant Transportation, 1769–1805." In *Science and Exploration in the Pacific: European Voyages to the Southern Oceans in the Eighteenth Century*, edited by Margarette Lincoln, 81–100. Woodbridge, England: Boydell Press with the National Maritime Museum, 1998.
Riley, Margaret. "The Club at the Temple Coffee House Revisited." *Archives of Natural History* 33, no. 1 (2006): 90–100.
Roberts, R. A., ed. *Manuscripts of The Earl of Egmont: Diary of the First Earl of Egmont (Viscount Percival)*. Vol. 2, 1734–1738. London: H. M. Stationary Office, 1923.
Robinson, Daniel F. *Confronting Biopiracy: Challenges, Cases and International Debates*. London: Earthscan, 2000.
Rodney, Walter. *A History of the Upper Guinea Coast, 1545 to 1800*. New York: Monthly Review Press, 1980.
Romans, Bernard. *A concise natural history of East and West Florida*. New York, 1775.
Roos, Anna Marie. *Web of Nature: Martin Lister (1639–1712), the First Arachnologist*. Leiden: Brill, 2011.
Rose, Edwin. "William Houstoun." In *The Collectors: Creating Hans Sloane's Extraordinary Herbarium*, edited by Mark Carine, 140–45. London: Natural History Museum, 2020.
Ryan, William J. "Imperfect Knowledge: Medicine, Slavery, and Silence in Hans Sloane's Philosophical Transactions and the 1721 London Pharmacopoeia." In *Drugs on the Page: Pharmacopoeias and Healing Knowledge in the Early Modern Atlantic World*, edited by Matthew James Crawford and Joseph M. Gabriel, 121–40. Pittsburgh: University of Pittsburgh Press, 2019.
Safier, Neil. *Measuring the New World: Enlightenment Science and South America*. Chicago: University of Chicago Press, 2008.
Salmon, Michael A. *The Aurelian Legacy: British Butterflies and Their Collectors*. Berkeley: University of California Press, 2000.
Sarr, Felwine, and Bénédicte Savoy. "The Restitution of African Cultural Heritage: Toward a New Relational Ethics." November 2018. http://restitutionreport2018.com/sarr_savoy_en.pdf.
Savitt, Todd. "The Use of Blacks for Medical Experimentation and Demonstration in the Old South." *Journal of Southern History* 48, no. 3 (1982): 331–48.
Schaffer, Simon. "Golden Means: Assay Instruments and the Geography of Precision in the Guinea Trade." In *Instruments, Travel and Science: Itineraries of Precision from the Seventeenth to the Twentieth Century*, edited by Marie-Noëlle Bourguet, Christian Licoppe, and Heinz Otto Sibum, 20–50. London: Routledge, 2002.
———. "Natural Philosophy and Public Spectacle in the Eighteenth Century." *History of Science* 21, no. 1 (1983): 1–43.
Schermerhorn, Calvin. *The Business of Slavery and the Rise of American Capitalism, 1815–1860*. New Haven, CT: Yale University Press, 2015.
Schiebinger, Londa. *Plants and Empire: Colonial Bioprospecting in the Atlantic World*. Cambridge, MA: Harvard University Press, 2004.

———. *Secret Cures of Slaves: People, Plants, and Medicine in the Eighteenth-Century Atlantic World*. Stanford, CA: Stanford University Press, 2017.

Schiebinger, Londa, and Claudia Swan, eds., *Colonial Botany: Science, Commerce, and Politics in the Early Modern World*. Philadelphia: University of Pennsylvania Press, 2005.

Schuessler, Jennifer. "What Should Museums Do with the Bones of the Enslaved?" *New York Times* (April 20, 2021). www.nytimes.com/2021/04/20/arts/design/museums-bones-smithsonian.html.

Sedgwick, Romney. *The House of Commons, 1715–1754*. 2 vols. New York: Oxford University Press for the History of Parliament Trust, 1970.

Seth, Suman. *Difference and Disease: Medicine, Race, and the Eighteenth-Century British Empire*. Cambridge, MA: Cambridge University Press, 2018.

Shapin, Steven. *A Social History of Truth: Civility and Science in Seventeenth-Century England*. Chicago: University of Chicago Press, 1995.

Sheridan, Richard B. "The Guinea Surgeons on the Middle Passage: The Provision of Medical Services in the British Slave Trade." *International Journal of African Historical Studies* 14, no. 4 (1981): 601–25.

Shiva, Vandana. "Bioprospecting as Sophisticated Biopiracy." *Signs* 32, no. 2 (Winter 2007): 307–13.

Shumway, Rebecca. *The Fante and the Transatlantic Slave Trade*. Rochester: University of Rochester Press, 2011.

Smallwood, Stephanie. *Saltwater Slavery: A Middle Passage from Africa to American Diaspora*. Cambridge: Harvard University Press, 2007.

Smeathman, Henry. *Plan of a Settlement to be made near Sierra Leona, on the Grain Coast of Africa*. London, 1786.

———. "Proposals for Printing by Subscription, Voyages and Travels in Africa and the West-Indies, from the year 1771, to the year 1779 inclusive." London, [1780].

———. "Some Account of the Termites, Which are Found in Africa and Other Hot Climates. In a Letter from Mr. Henry Smeathman, of Clement's Inn, to Sir Joseph Banks, Bart. P. R. S." *Philosophical Transactions of the Royal Society of London* 71 (1781): 139–92.

Smith, Charles Hamilton. "Memoir of Dru Drury." In *Mammalia*, edited by William Jardine, 17–71. Vol. 15 of *The Naturalist's Library*. London, 1842.

Soelberg, J., A. Asase, G. Akwetey, and A. K. Jäger. "Historical Versus Contemporary Medicinal Plant Uses in Ghana." *Journal of Ethnopharmacology* 160 (2015): 109–32.

Stearn, William T. "Philip Miller and the Plants from the Chelsea Physic Garden Presented to the Royal Society of London, 1723–1796." *Transactions of the Botanical Society of Edinburgh* 41, no. 3 (1972): 293–307.

Stearns, Raymond Phineas. "James Petiver: Promoter of Natural Science, c. 1663–1718." *Proceedings of the American Antiquarian Society* 62, pt. 2 (October 1952): 243–358.

———. *Science in the British Colonies of America*. Urbana: University of Illinois Press, 1970.

Stewart, Larry. "The Edge of Utility: Slaves and Smallpox in the Early Eighteenth Century." *Medical History* 29, no. 1 (1985): 54–70.

———. "Global Pillage: Science, Commerce, and Empire." In *The Cambridge History of Science*. Vol. 4, *Eighteenth-Century Science*, edited by Roy Porter, 825–44. Cambridge: Cambridge University Press, 2003.

———. "Public Lectures and Private Patronage in Newtonian England." *Isis* 77, no. 1 (March 1986): 47–58.

———. *The Rise of Public Science: Rhetoric, Technology, and Natural Philosophy in Newtonian Britain, 1660–1750*. Cambridge: Cambridge University Press, 1992.

Stewart, Mart A. "'Policies of Nature and Vegetables': Hugh Anderson, the Georgia Experiment, and the Political Use of Natural Philosophy." *Georgia Historical Quarterly* 77, no. 3 (Fall 1993): 473–96.

Strong, L. A. G. *Dr. Quicksilver, 1660–1742: The Life and Times of Thomas Dover, M.D.* London: Andrew Melrose, 1955.

Superb Collection of Insects, Elegant Cabinets, &c, A Catalogue of the Most Capital Assemblage of Insects Probably ever Offered to Public Sale; Consisting of upwards of Eleven Thousand different Specimens, Collected from all the Countries with which Great Britain has any intercourse . . . by Mr. Dru Drury. London, 1805.

Swann, Marjorie. *Curiosities and Texts: The Culture of Collecting in Early Modern England*. Philadelphia: University of Pennsylvania Press, 2001.

Sweet, James H. *Domingos Álvares, African Healing, and the Intellectual History of the Atlantic World*. Chapel Hill: University of North Carolina Press, 2011.

Sweet, Julie Anne. "A Misguided Mistake: The Trustees' Public Garden in Savannah, Georgia." *Georgia Historical Quarterly* 93, no. 1 (Spring 2009): 1–29.

Taylor, R. "Sea-Surgeons and the Company of Barbers and Surgeons." *Journal of the Royal Naval Medical Service* 87, no. 2 (2001): 98–103.

Te Heesen, Anke. "Accounting for the Natural World: Double-Entry Bookkeeping in the Field." In *Colonial Botany: Science, Commerce, and Politics in the Early Modern World*, edited by Londa Schiebinger and Claudia Swan, 237–51. Philadelphia: University of Pennsylvania Press, 2005.

Terrall, Mary. "African Indigo in the French Atlantic: Michel Adanson's Encounter with Senegal." *Isis* 114, no. 1 (March 2023): 2–24.

———. *Catching Nature in the Act: Réaumur and the Practice of Natural History in the Eighteenth Century*. Chicago: University of Chicago Press, 2014.

Thornton, John. *Africa and Africans in the Making of the Atlantic World, 1400–1680*. Cambridge: Cambridge University Press, 1992.

Tobin, Beth Fowkes. *The Duchess's Shells: Natural History Collecting in the Age of Cook's Voyages*. New Haven, CT: Yale University Press, 2014.

Walker, Geoffrey J. *Spanish Politics and Imperial Trade, 1700–1789*. Bloomington: Indiana University Press, 1979.

Walker, Timothy D. "The Medicines Trade in the Portuguese Atlantic World: Acquisition and Dissemination of Healing Knowledge from Brazil (c. 1580–1800)." *Social History of Medicine* 26, no. 3 (August 2013): 403–31.

Wallis, Patrick. "Exotic Drugs and English Medicine: England's Drug Trade, c. 1550–c. 1800." *Social History of Medicine* 25, no. 1 (2011): 20–46.

Warsh, Molly. *American Baroque: Pearls and the Nature of Empire 1492–1700*. Chapel Hill: University of North Carolina Press, 2018.

Watson, W. N. Boog. "The Guinea Trade and Some of Its Surgeons (with Special Reference to the Royal College of Surgeons of Edinburgh)." *Journal of the Royal College of Surgeons of Edinburgh* 14, no. 4 (1969): 203–14.

Weaver, Karol K. *Medical Revolutionaries: The Enslaved Healers of Eighteenth-Century Saint Domingue*. Urbana: University of Illinois Press, 2006.

Webster, Jane. "The Material Culture of Slave Shipping." In *Representing Slavery: Art, Artefacts and Archives in the Collections of the National Maritime Museum*, edited by Douglas Hamilton and Robert J. Blyth, 104–17. London: Lund Humphries, 2007.

Westwood, J. O. *Illustrations of Exotic Entomology, Containing Upwards of Six Hundred and Fifty Figures and Descriptions of Foreign Insects, interspersed with Remarks and Reflections on their Nature and Properties. By Dru Drury. A New Edition* . . . 3 vols. London, 1837.

Wilder, Craig Steven. *Ebony and Ivy: Race, Slavery, and the Troubled History of America's Universities*. New York: Bloomsbury, 2013.

Williams, Eric. *Capitalism and Slavery*. Chapel Hill: University of North Carolina Press, 1944.

Winterbottom, Anna. *Hybrid Knowledge in the Early East India Company World*. Basingstoke, Hampshire: Palgrave Macmillan, 2016.

Wood, Betty. "The Earl of Egmont and the Georgia Colony." In *Forty Years of Diversity: Essays on Colonial Georgia*, edited by Harvey H. Jackson and Phinizy Spalding, 80–96. Athens: University of Georgia Press, 1984.

Wulf, Andrea. *The Brother Gardeners: Botany, Empire and the Birth of an Obsession*. New York: Knopf, 2009.

Zahedieh, Nuala. "The Merchants of Port Royal, Jamaica, and the Spanish Contraband Trade, 1655–1692." *William and Mary Quarterly* 43, no. 4 (October 1986): 570–93.

INDEX

Page numbers in italics refer to illustrations.

abolitionism, 5, 9, 159, 167, 177, 178, 179, 183
Adanson, Michel, 158, 159
Addoe, William, 161
Africa (slave ship), 140, 143, 165, 176–77
African natural knowledge, 43–46, 49–51, 59–68, 103, 172–73
Aird, Robert, 153, 155
Angola. *See* West Central Africa
Anomabu, 2, 33, 141
antislavery, 5, 9, 159, 167, 177, 178, 179, 183
asiento (slaving contract), 4, 71–99, 102–3, 110–11, 115, 121–23, 208n52. *See also* South Sea Company
Assiento (slave ship), 94–95, 104–7, 204n8, 205n13. *See also* Houstoun, William
assintee, 67–68, 199n67
Atlantic World, 8, 11, 13, 35, 72, 99, 104, 184, 188n16

Bags, Captain, 87
Baldwyn, Ambrose, 59–60
Banana Islands, 152, 160–62, 165, 171, 173, 213n23, 214n30
Bance Island, 152, 163–64, 166
Banks, Joseph, 98, 157, 174, 178, 182, 204n65

Barbados, 33
Barber, Miles, 165, 176, 216n74
Barcklay, Robert, 19–21
Barlow, Samuel, 58, 67, 199n65
Bartar, Edward, 24, 25–30, 34–35, 41, 191n28
Berlin, Anders, 165, 173–74, 177, 178, 214n25, 214n36
Bight of Benin, 30, 34, 53, 59, 67
Bight of Biafra, 128, 130, 144
bioprospecting, 69, 101–23, 206n23
botany, 28, 38, 52, 95, 97, 101–23; economic botany, 113–14. *See also* collections: botanical
Boyle, Robert, 46
British Museum, 4, 33, 98
Brown, William, 35
Brydges, James. *See* Chandos, 1st Duke of
Buenos Aires, 2, 76, 79, 81, 88, 89, 90, 93, 200n17
Burnet, John, 1–5, 12–14, 71–72, 81–87, 90–92, 94, 99, 189n30, 201n21

Calabar, 128, 130, 137, 145–47
Campbell, Mr., 32–33
Campeche, 94, 107–8

235

Cape Coast Castle, 21–28, 24, 31, 34–35, 43–44, 49, 53–58, 61–69
Caribbean, 58, 66–68, 81, 86, 91–92, 179, 189n23, 191n22, 216n67
Cartagena, 34, 81–85, 116–19, 122, 207n46
Catesby, Mark, 106, 113, 157, 206n29
Cavendish Bentinck, Margaret. *See* Portland, Duchess of
Chandos, 1st Duke of (James Brydges), 49–69, 198n58, 199n67
Chelsea Physic Garden, 97, 110, 122, 191n18, 206n30, 207n46
cinchona, 56, 63, 73, 88, 118, 121
Clarendon (slave ship), 49–51, 56, 60–61
Cleland, Mr., 137
Cleveland, James, 160–62, 213n22, 213n23
Cleveland, John, 160–62, 171, 213n22
collecting, 6; and commerce, 6–8, 25, 47, 65, 88, 91, 123, 125–26, 138, 166, 178; directions for, 60, 90, 127, 132, 134–37, 140, 145, 149, 211n51; from a slave ship, 1–6, 12–13, 15, 30–31, 33–39, 103–10, 135–37, 155, 166; supplies for, 35, 127, 129, 132–45, 150, 155, 157. *See also* collections; collectors
collections: botanical, 28, 41–44, 49, 60–61, 89–90, 93, 95–97, 113, 117, 174, 182; by John Burnet, 82–84; catalogs of, 43–46, 60–61, 69, 95, 128, 146; *Clarendon* collection, 49–50, 60–62, 69; as commodities, 127, 134, 140, 153–55, 157, 175; entomological, 125–70, 174, 181–82; man-made, 33; marketplace for, 4, 36–37, 157–58; mineral, 93; as objects of experimental study, 170–71; preservation of, 174–75; surgical and anatomical, 14; and taxonomy, 170; zoological, 2–3, 31, 33, 34, 35, 93
collectors, 6–7, 133; of African descent, 11, 15, 21, 24, 27–28, 37, 55, 83, 106, 123, 128–31, 135–36, 145–46, 161; of Indigenous descent, 83, 106, 108, 110, 123; traveling naturalists, 153–70. *See also* slave ship captains; slave ship carpenters, as collectors; slave ship mates, as collectors; slave ship surgeons;

slaving agents; slaving factories: factory surgeons
colonialism: and medical knowledge, 11, 14, 22, 44–47, 59–64, 88, 183; and natural history, 52, 63, 69, 77, 93, 102, 111–14, 158–59, 183–86
Company of Merchants Trading to Africa, 127, 130
contraband. *See* smuggling
contrayerva (drug), 107–8, 109, 205n18
Corker, Charles, 161
Costa, Emanuel Mendes da, 129
Coward, Edward Noll, 86–87
Cowley, Richard, 144, 149
Cowley, Stretch, 144
Crawford, David, 35, 36

Dillenius, Johann Jacob, 41, 94; *Hortus Elthamensis*, 94
direct trade, between Britain and Africa, 52, 139, 167
Dolben's Act, 38
Douglas, James, 2, 3, 33, 82, 94
Dover, Thomas, 76–79, 81, 88–90, 200n17
drugs, 63, 92; African, 43–46, 49, 54–56, 59–62, 64–68; bioprospecting for, 2, 13, 51, 53–57, 59, 63, 64, 66, 73, 93, 102, 114, 118–19; market for, 44, 51, 55, 62, 67, 93; and the slave trade, 55, 62–63, 67; Spanish American, 82–83, 107–10, 114, 118–19, 205n19; specifics, 62–63
Drury, Dru, 16, 125–51, 156–58, 167, 173–74; "Directions for Collecting," 134–37, 140, 145, 149; *Illustrations of Natural History*, 125, 129, 131, 138, 146, 147, 149–51, 174, 209n15, 212n64; museum, 149
Drysale, Mr., 140–42
dyes, 2, 4, 13, 53, 65, 67, 73, 82–83, 92–93, 111

embasnobah, 60–61
enslaved collectors. *See* collectors

Fabricius, Johan Christian, 182
Fante people, 22–24, 30, 40, 43, 46, 49–50, 60–61
Fothergill, John, 156–57, 167–68, 177–78, 182
Fraser, James, 30–32, 36, 38, 193n49

Gabon Estuary, 16, 124, 125–30, 148, 151
Gambia, 19–20, 24, 40, 53, 58, 66–67
geographic knowledge, 11, 64, 69, 77–78, 85, 198n59
Georgia, 16, 102, 110–16, 120–21, 206n30
Gold Coast 21–29, 41, 43–46, 49–50, 60–62, 143
Goliath beetle (*Goliathus goliatus*), 16, *124*, 125–51, *126*, 209n10
Gower, William, 49
Grosvenor, Seth, 66
Guinea. *See* West Africa

Hall, Francis, 90, 91, 93–94
Halley, Edmond, 84–85
Harris, Moses, 129, 131, *133*, 212n64
Haslewood, Mr., 86–87
Havana, 94, 105, 119
herbaria, 22, 25, 29, 41, 44, 45, 90, 94, 95, 97–98, 182. *See also* Sloane Herbarium
Herle, Digory, 1–4
Herrera, Juan de, 84–85
Hill, David, 160, 165–66, 174, 214n36
Hoskins, John, 92
Hough, Mr., 135–36
Hound (Royal Navy), 130
Houstoun, James, 117, 203n56
Houstoun, William, 16, 92, 94–98, 101–23, 182–83, 203n56, 204n8, 205n13, 206n29, 206n30; "An Account of the Contrayerva," 108–9; "Catalogus Plantarum in America Observatarum," 95, *96*, 98–104, 203n50; *Reliquae Houstounianae*, 204n65
Hunter, John, 174, 216n66
Hunter, William, 129, 151, 174, *180*, 181–83, 186, 215–16n66
Hunterian Museum, 17, 151, 181–83, 215–16n66

Îles de Los, 152, 163, 165, 176
ipecacuanha, 31, 94, 117–19, 121, 208n54

jalap (*Ipomoea purga*), 108, 110, 205n19, 205n20

Jamaica, 19–20, 35, 67–68, 86–87, 91–95, 103–8, 110, 115–18, 199n67, 203n50
James, Robert. *See* Petre, 8th Baron
James, William, 168, 169
Jesson, George, 37, 89–91, 93, 98
Johnson, Thomas, 19

Kirckwood, John, 13–14, 41

Lee, James, 156–57
Legacies of British Slave-ownership Project, 9, 10
Lightfoot, Robert, 33
Linnaeus, Carolus, 6, 41–42, 97, 167, 173, 178
Lister, Martin, 41
Liverpool, 128, 143, 145–48, 159, 168, 176, 211n39
Lynn, Francis, 66

Male, Thomas, 130
mariners, 37; collecting for Drury, 127–51; and history of science, 11, 104; slaving mariners, 8, 11, 13, 15, 20, 21, 31–41, 91, 126, 159, 162, 166–67, 212n65
Martyn, John, 97; *Historia plantarum rariorum*, 97
materia medica, 31, 44–45, 195n75
Maupertuis, Pierre-Louis Moreau de, 63
mercantilism, 73, 111
Millar, Robert, 16, 102, 115–23, 207n46, 208n54
Miller, Philip, 97–98, 112–13, 122, 203n56, 207n46; *Gardeners Dictionary*, 97, 112, 122; herbarium, 97–98, 122
Miller, Thomas, 208n10
Moze, Paul, 57
museums: and colonialism, 14, 183–86, 217n8, 217n9; and slave trade, 6, 17, 28, 32–33, 44, 47, 94, 98, 181–86, 189n22
Mylam, Mr., 93–94

natural commodities: and Georgia, 113–14, 121; and mercantilism, 73, 111; searching for, 16, 93, 113; and slaving companies, 13, 23, 51–59, 63–69, 92; Spanish American, 71, 73, 123; West African, 127, 139, 156

Natural History Museum (London), 17, 28, 44, 98, 189n22, 218n10
New Providence (slave ship), 19, 21

Ogilvie, David, 125, 128, 129, 151
Oglethorpe, James Edward, 206n30
Oldenburg, Henry, 25
Oxford University, 93–94; Oxford University Herbaria, 17, 94

Parke, Nonus, 128–30, 135–37, 145–48, 209n7, 212n57
Partington, Henry, 76, 79
patronage, 16, 157; of John Burnet, 85–86; of William Houstoun, 85, 112–15, 121–23, 206n29, 206n30; of Robert Millar, 118, 120; of Henry Smeathman, 157, 159, 169, 174, 178, 182
Patton, David, 89
Pennant, Thomas, 166
Petiver, James, 4, 25; and Edward Bartar, 25–28, 35; "Catalogue of some *Guinea-Plants*," 43–46, 194n70, 195n73; and collectors of African descent, 106, 191n22, 193n55; *Gazophylacii Naturae & Artis*, 14, 31–32, 36, 41, 42; *Musei Petiverani*, 36; museum, 4, 25, 34, 43, 44, 46, 190n4, 191n28, 195n74; natural historical images, 41, 194n68; reliance on slave trade, 14, 19–22, 25, 30–40, 46, 89, 93, 136, 192n40; *The South-Sea Herbal*, 88
Petre, 8th Baron (Robert James), 98, 118, 122, 206n30
Phipps, James, 54, 58, 62
Pintado, Manuel Lopéz, 119–20
Planer, Richard, 33–36, 41
Plukenet, Leonard, 20, 26, 41, 191n18
Portland, Duchess of (Margaret Cavendish Bentinck), 157, 182
Portobelo, 4, 71, 81, 86–87, 91, 106, 112, 118

Royal African Company, 15–16, 21–35, 47, 49–70, 127, 195n78
Royal Society of London, 2, 28, 33, 36, 46, 52, 56, 84–85, 118, 120; collections, 24, 94; meetings, 20, 22, 25, 33, 40, 97, 103, 108, 112; membership, 27, 41, 43, 103, 108,

114–15, 118; *Philosophical Transactions*, 43, 74, 85, 88, 94, 108

São Tomé, 2, 12, 27, 33, 130
Senegambia, 127, 158–59, 213n19
Sherard, William, 20, 41, 88, 90, 93–94
Sierra Leone, 16, 57–58, 68, 143–44, 152, 153–70, 181–83, 212n5
Skeen, James, 33–35, 192n41
slave ship, 12, 14, 79; descriptions of, 165, 176–78, 179; itineraries and collecting patterns, 34, 37; and medical knowledge, 55, 67; transporting specimens, 21, 49–51, 61, 87, 91, 106, 140, 155, 160, 167, 181; transporting traveling naturalists, 57, 164–66. *See also* collecting: from a slave ship
slave ship captains, 12, 31–33, 58, 90, 153–54, 159, 166; as collectors, 33, 89, 91, 130, 132, 135, 144, 146
slave ship carpenters, as collectors, 140–42
slave ship mariners: as go-betweens, 34, 35, 39, 91; natural knowledge of, 66–67
slave ship mates, as collectors, 128, 130, 137, 140, 142–44, 148–49
slave ship surgeons, 8, 13, 194n59; as collectors, 1–4, 12–14, 19–20, 28, 30–33, 37–39, 41, 81, 95–97, 103–10, 135–36, 182
slave trade, British transatlantic, 5, 8, 11, 18, 21–24, 32–34, 47, 50, 69–72, 102–4, 126–27, 139, 159, 183; bulking centers, 163–64; and geographies of collecting, 24, 31–35, 64–68, 90–92, 111–12, 122–23, 148, 166; historiography of, 8–12, 188n16; impact on natural historical collecting, 5, 8, 12–17, 21, 28–30, 87–91, 98–99, 135, 151, 155, 160–62, 167–68; natural historical profits 1, 11, 20, 40–47, 71–99, 121–22, 149–50, 183–85; and the Royal Navy, 76, 127, 130, 139; within Sierra Leone, 155, 162–63
slave trade, Spanish transatlantic, 71–72, 74–75, 105–6, 123, 204n3
slaving agents: as collectors, 25–30, 41, 53, 56, 58; facilitating transportation, 91–92, 112, 115–18, 163–68; and natural knowledge, 25, 51, 54, 59–68

238   Index

slaving companies, and natural history, 8, 13, 50–69, 71–99, 101–12, 115

slaving factories, 23, 69, 75, 106, 163–64; collecting in the vicinity of, 53–55, 58, 63, 65–68, 93, 116–19; factory surgeons, 57, 71–72, 81–92; plans for plantations, 57–59, 68

Sloane, Hans: collections, 4, 32–33, 37, 82–86, 89, 91, 95, 122, 201n21, 205n13; as patron, 101, 104, 112–13, 115, 117; and Royal African Company, 56–69; and Royal Society of London, 40, 108; *Vegetable and Vegetable Substances*, 60–61, 69, 95. See also Sloane Herbarium

Sloane Herbarium, 28, 44, 45, 95, 190n7, 191n28. See also Sloane, Hans

Smeathman, Henry, 16, 152, 153–70, 181–82, 212n62; African assistants, 171–73; collections, 173–75; *Plan of a Settlement to be made near Sierra Leone*, 179, 212n5; "Some Account of the Termites, Which are Found in Africa and Other Hot Climates," 169–73, 170, 172

Smeathman, John, 168

smuggling, 12, 74–75, 86, 99, 103; of plants, 101–23

Smyth, John, 21–22, 27, 31, 39–41, 43–46, 49

South Sea Company, 71–72, 77, 85, 86, 189n30; *asiento* trade, 4, 75–76, 79, 102; facilitating collecting, 1, 16, 72, 81–84, 86–94, 99, 101–23; and natural commodities, 93. See also *asiento*

Spanish America, 1–2, 16, 71–123, 204n3

specifics. See drugs

specimens. See collections

surgeons. See slave ship surgeons; slaving factories: factory surgeons

Sutro Library (San Francisco), 98, 182, 183

Tinker, John, 49–50, 61

Toller, William, 76–80, 89, 93, 98, 200n10; "The History of a Voyage to the River of Plate & Buenos Ayres from England," 76–80

Trustees of Georgia, 113–15, 119, 120–22, 206n30, 208n54

Upper Guinea Coast, 153–79

Venus (slave ship), 140–42

Veracruz, 89, 92, 94–95, 101, 105–8, 110, 119–21

visual knowledge, 40, 130–32

Walker, William, 33

Wallace, William, 140, 143–44, 211n51

Warwick (Royal Navy), 76–81, 89, 200n17

Watts, William, 33

West Africa, 15–16, 19–70, 124, 127–51, 153–79, 190n4, 190n11, 195n78

West Central Africa, 15–16, 19–47, 105, 124, 127–51, 153–79, 190n4

Westcomb, Daniel, 9

West Indies. See Caribbean

Westwood, John Obadiah, 150

Whydah, 30, 31, 33, 53, 56, 58, 59, 60, 67, 86

Wilding, Captain, 165

Williams, Eric, 188n20, 189n23

Williams, Thomas, 130, 132, 144, 148

Wiltshire (slave ship), 1–4, 13–14, 81, 189n30; voyage of, *xvi*

Index

www.ingramcontent.com/pod-product-compliance
Lightning Source LLC
Chambersburg PA
CBHW021854230426
43671CB00006B/384